BOOK **1** IN THE TRAILS BY

Natural Surface Trails *by* Design

Physical and Human Design Essentials of Sustainable, Enjoyable Trails

TROY SCOTT PARKER

Natureshape

BOULDER, COLORADO, USA

Natureshape

Natureshape LLC
8285 Kincross Drive
Boulder, CO 80301-4228 USA
Ph (303) 530-1785
Fax (303) 530-4757
www.natureshape.com
info@natureshape.com

No part of this book may be reproduced or transmitted in any form or by any means, electronic or mechanical, including photocopying and recording, or by an information storage and retrieval system, without permission in writing from the publisher.

ISBN: 0-9755872-0-X
Library of Congress Control Number: 2004107571

Text, photographs, and illustrations copyright © 2004 by Troy Scott Parker
All rights reserved

Cover design, book design, layout, typography, illustrations, and photo prepress by Natureshape LLC. *In addition to publishing and trail consulting, Natureshape also provides award-winning book cover and interior design, book layout, graphic design, typesetting, photo prepress, and website design services to other publishers and entities for any type of book or project. Please visit www.natureshape.com or contact Natureshape for more information.*

A catalog of titles is available on the Natureshape website at www.natureshape.com. Also, see the last page of this book.

Use of the information in this book is at the sole risk of the user. Users should evaluate the applicability of any recommendation in light of particular situations and changing practices.

Printed in South Korea

10 9 8 7 6 5 4 3 2 1

Dedicated to architect Christopher Alexander and his groundbreaking books,

The Timeless Way of Building and *A Pattern Language,*

for leading me toward more effective ways to comprehend natural surface trails;

and to the countless individuals and agencies

who shaped the many trails from which I've learned so much

Books in the Trails by Design Series

Natural Surface Trails by Design:
Physical and Human Design Essentials of Sustainable, Enjoyable Trails

Shaping Natural Surface Trails by Design:
Key Patterns for Forming Sustainable, Enjoyable Trails

Managing Trails by Design:
Integrating Stewardship, Sustainability and the Trail Experience

It is said there are three types of people:
   1. *Those who make things happen*
   2. *Those who watch things happen*
   3. *Those who wonder what happened*

Contents

1 A Fundamentally Different Approach to Trail Design 7
    The Old Problems 7
    A Different Approach 7
    A Single, Flexible System of Thought 8

2 Introducing the Foundation Level of Basic Forces & Relationships 11

3 Human Perception 13
    Natural Shapes 13
    Anchors 16
    Edges 17
    • Following, Approaching, and Crossing Edges 19
    Gateways 20
    The Power of Combinations 21
    Conclusion 22

4 Human Feelings 23
    Safety 24
    Efficiency 25
    Playfulness 27
    Harmony 28
    • How Our Modality Affects Harmony 31
    Human Perception, Feelings, and Stewardship 31
    • Example: How Perception and Feelings Affect Switchbacks 32

5 Physical Forces 35
    Compaction 35
    Displacement 37
    How Compaction and Displacement Interact 38
    • Compaction & Displacement by Modality 41
    Erosion 42
    How Erosion, Compaction, and Displacement Interact 43
    Physical Forces in Context 44

6 Tread Materials 45
- Tread Texture 45
- ▪ Textures and Behaviors of Common Tread Materials 46
- Tread Texture and Tread Performance 48
- ▪ A Simple Field Test for Identifying Clay, Silt, Sand, and Loam 48
- Tread Hardening 49
- Testing Performance of Tread Materials 49
- ▪ How To Estimate the Performance of Any Tread Material 50

7 Tread Watersheds 51
- Tread Watershed 51
- Tread Watershed Factors 53
- Tread Watershed Size 54
- Watershed Slope 54
- Runoff Potential 55
- Splash Erosion 55
- Tread Width 56
- Weather, Climate, and Microclimate 56
- Water Sources 56
- Tread Texture 57
- Trail Use (Compaction and Displacement) 57
- Tread Grade and Tread Length 58
- Dip Sustainability 60
- Putting All the Factors Together 62
- Predicting the Future 62
- ▪ Quick Reference: Tread Erosion and Water Damage Risk 62

8 Trail Evaluation: Reading Trails Like a Book 63
- Old Problems with Trail Evaluation 63
- Easier Evaluation with the Foundation Level 63
- ▪ Quick Trail Evaluation Form—Foundation Level 64
- Evaluation as a Trail Design Exercise 65
- Trail Evaluations Around the Nation 66

9 Trails by Design 75
- Using the Foundation Level 75
- ▪ The Foundation Level in a Nutshell 76
- The Next Two Levels 77

About the Author 79

● CHAPTER 1

A Fundamentally Different Approach to Trail Design

Nothing is more dangerous than an idea, when you have only one idea.
– EMILE CHARTIER, French philosopher

The Old Problems

Traditional attempts to explain trail design typically encounter the same problems:

- While some individuals are skilled trail designers, most of them cannot tell you how they do it because there's no communication tool. We don't even have words for basic aspects of trails.
- Trail design is often approached as a collection of "standards" for trail width, grade, turning radius, clearance, specific construction techniques, and trail structures. However,
    - different trail types, uses, materials, grades, slopes, and travel speeds all require different standards, as do different combinations of these; and
    - the needs and purposes of visitors, land managers, and trails themselves are far richer and more diverse than limited standards can accommodate.
- The visitor's trail experience—the major reason for having recreational trails—is often shortchanged or ignored, especially when limited standards dictate design.
- Some trail information sources prescribe techniques that assume that trail surfaces are static like concrete when in fact they change shape all the time from trail use, erosion, and more. If their shape changes enough, the original prescription may be inappropriate. *But they don't tell you that.*
- Some individuals think of "fighting" trail erosion. That's like fighting nature itself. It tends to lead to engineered solutions when we can often limit erosion in more naturalistic ways.
- Evaluating trails has traditionally been difficult because we couldn't define, describe, or measure them in any concise way.

All of these are instances of having "only one idea" and seeing trails from that little window, *thinking that's all there is.*

A Different Approach

What if we had a fundamentally different approach for trails, an approach where the "what, how, and where" flow from the key question, "why"?

- What if we could work with trails as complex yet manageable systems of forces and relationships, of organized causes and consequences?
- What if we could better understand of the scores of human and physical particulars of trail uses, types, materials, desired trail experiences, and construction techniques while encouraging and enhancing their richness, diversity, customization, sustainability, and enjoyment?
- What if we had a system of thought that teaches us to recognize all of the underlying forces and relationships in each unique situation, helps predict the future of each situation, and directly generates appropriate trail solutions? And what if the system also tells us what *won't* work in a particular situation?
- What if we had a simple language to facilitate trail design, as well as for learning and teaching it?
- What if trail evaluation were quick and easy, without cumbersome methods, and with consistent results by different evaluators?
- What if the system helps increase respect, appreciation, and stewardship for natural resources in both practitioners and trail visitors?
- And what if all of this were relatively quick and easy to learn and use for both novices and experienced trail designers?

A Single, Flexible System of Thought

The system exists. It efficiently leads to the same types of outcomes achieved by skilled trail designers. And because most of its pieces have been under our noses forever, it's relatively easy to learn.

It's complex and flexible enough to work in any and all trail contexts, yet simple and concrete enough to be used directly—even intuitively—by anyone who takes the time to learn and practice it, from novices to seasoned designers. It explains what happens in the real world, and why. In manageable detail, it tells us what to look for, how to use what we see, how to combine variables, how to predict future behavior, and how to work with cause and effect.

Based on human behavior and physics, it's a system of thought that goes to the heart of the "whys" of all trail situations rather than stopping at the "whats." *And knowing the whys inherently suggests what to do at each step.* It enables us to quickly understand new or complex situations in terms of an organized set of human and physical forces and relationships.

In short, the system teaches the broad strokes and details of sustainable trail design in simple, flexible, and manageable terms. It gets into the heads of both trail visitors and trail designers while using what nature, the site, materials, and management can offer. It's internally consistent yet expandable and open-ended. It makes it easy to evaluate trails, both existing and proposed, and it makes it much easier for a group of people to reach consensus on trail design issues. And—if you'll let it—it helps produce trails that are as naturalistic, sustainable, and wonderful as they can be given their context.

The system is based on shapes, forces, patterns, relationships, and language interacting across three levels. Most aspects of the higher levels are explicitly driven by the lower levels, making it primarily a bottom-up system.

The three levels are:

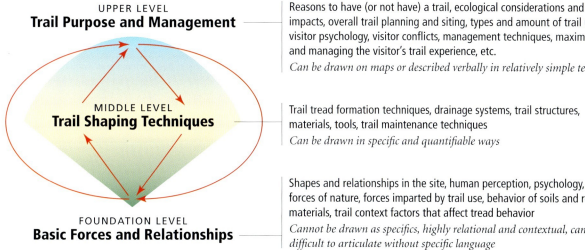

UPPER LEVEL — Trail Purpose and Management
Reasons to have (or not have) a trail, ecological considerations and impacts, overall trail planning and siting, types and amount of trail use, visitor psychology, visitor conflicts, management techniques, maximizing and managing the visitor's trail experience, etc.
Can be drawn on maps or described verbally in relatively simple terms

MIDDLE LEVEL — Trail Shaping Techniques
Trail tread formation techniques, drainage systems, trail structures, materials, tools, trail maintenance techniques
Can be drawn in specific and quantifiable ways

FOUNDATION LEVEL — Basic Forces and Relationships
Shapes and relationships in the site, human perception, psychology, forces of nature, forces imparted by trail use, behavior of soils and rocky materials, trail context factors that affect tread behavior
Cannot be drawn as specifics, highly relational and contextual, can be difficult to articulate without specific language

Starting from the bottom:

Foundation Level: Basic Forces and Relationships

The Foundation Level consists of core concepts, forces, and relationships of psychology and physics that directly support all aspects of natural surface trails. Sustainability begins here. *Wherever any of the factors or forces at this level are not accommodated in the middle or top levels, there **will** be problems in trail construction, trail maintenance, or management.* Graphically and literally, these basic forces and relationships—the lowest tip of the cone—support the wide variety of trail shaping techniques at the Middle Level.

Info source: The next chapter introduces the Foundation Level, and the remainder of this book discusses it in detail.

Middle Level: Trail Shaping Techniques

Natural surface trail design and construction techniques—outslope, backslope, rolling grade, drainage dips, tread* hardening, and more—are sometimes treated as if they are the root of trail design. But they're not. Foundation Level physics and psychology underlie them.

In this tri-level structure, all construction techniques—traditional and new—are on the Middle Level based on sustainable combinations of Foundation Level forces and relationships. Trail shaping (design, construction, maintenance) techniques gain depth, sustainability, richness, and ability to engender stewardship when explicitly grown from their real roots in physics and psychology. In addition, some Middle Level techniques based on the Foundation Level can shape more sustainable, naturalistic trails with less initial expense and lower maintenance requirements than traditionally constructed trails.

* "Tread" is the actual surface on which we actively place feet, hooves, wheels, etc. In this book, "tread" is used to describe the physical travel surface while "trail" refers to the broad concept of a trail as a continuous travelway.

Middle Level techniques use Foundation Level concepts to:

- Improve sustainability and reduce maintenance through design.
- Help us find (or develop) the most appropriate solution for each actual underlying problem.
- Improve traditional trail design and construction techniques.
- Improve effectiveness of trail maintenance.
- Reduce shaping costs where feasible.
- Improve the visitor's trail experience.
- Use trail shaping techniques to better serve Upper Level trail purpose and management.

Info source: Book 2 in the Trails by Design series, *Shaping Natural Surface Trails by Design: Key Patterns for Forming Sustainable, Enjoyable Trails,* discusses the key Middle Level patterns that optimize most aspects of natural surface trails. These patterns are "grown from" basic forces and relationships in the critical Foundation Level.

Upper Level: Trail Purpose and Management

The Upper Level addresses larger-scale trail planning and management issues including ecological and social impacts of trails, whether or not a trail should even exist, what a trail should visit or avoid, trail corridor selection and planning, types and amount of trail use, visitor psychology, the intended visitor experience, single versus multiple use, management techniques, and comprehensive trail evaluation.

Graphically, the Upper Level is shown as a dome supported by trail shaping techniques. This is an accurate metaphor. The best-conceived and best-managed trails (at the top of the dome) are completely supported by trail shaping techniques (the Middle Level) and basic forces and relationships (the Foundation Level). Some Foundation Level concepts are used directly for Upper Level aspects such as overall corridor and trail alignment, managing the desired trail purpose, and providing the desired recreational experience.

This vertically integrated system tends to generate an Upper Level in harmony with (fully supported by) the Middle and Foundation Levels. And where conflicting goals or aspects exist at the top—as often occurs—the system reveals the conflicts and makes it easier to know the consequences of various alternatives and compromises.

On the other hand, unresolved forces in the Foundation and Middle Levels can destabilize trail management and/or work against the trail purpose. The tri-level structure reveals these problems, too—and can pinpoint what they are. If problems can be prevented or resolved at the lower levels, it won't be necessary to make purely administrative (and often ineffective) attempts to deal with them at the Upper Level.

Info source: Book 3 in the series, *Managing Trails by Design: Integrating Stewardship, Sustainability and the Trail Experience,* discusses Upper Level trail purpose and management as discussed above.

● CHAPTER 2

Introducing the Foundation Level of Basic Forces & Relationships

People only see what they are prepared to see.
  – RALPH WALDO EMERSON, American philosopher

Each and every natural surface trail is a part of nature subject to a range of human and natural forces that we cannot fully control. We know that. Each trail both creates and is affected by an entire web of relationships between its site, visitors, alignment, soils and materials, water, management, and far more. We know that, too. And the forces affect the relationships, and vice versa. So—*despite what trail standards may say or what the public, the land managers, or their bosses may want*—what **actually** happens on every piece of every trail is determined by basic forces and relationships acting relentlessly day after day, year after year.

Hence the way to truly understand trails is through their basic forces and relationships. But what are they?

Undoubtedly, some of them are at least partially familiar to you. Others, however, differ from what many of us were told or were prepared to see, and so we didn't see them. Yet they all seem obvious once they have been pointed out, given names, and organized into a single, logical structure that tells us where and how to look.

The Foundation Level *is* that single, logical structure. In eleven core concepts, it captures most of the human and physical forces and relationships that underlie natural surface trails. Only eleven—and they're all described in this thin little book.

As an introduction, the eleven concepts (italics below) are grouped into five categories (bold below):

Human Perception
How we perceive nature and trails, and how their shape affects us:
- *Natural shape* describes the basic shape which shapes all things in nature, including trails, and how we respond to that shape.
- *Anchors,* and their particular instances of edges and gateways, describe how we perceive landscapes and sites, as well as how trails can use anchors to feel like they belong in a site.

Human Feelings
How the trail affects our trail experience, and how our feelings affect the trail itself:
- *Safety* is our perceived level of safety—our comfort zone—for being on a trail.
- *Efficiency* is our desire that the trail be easier to use than to bypass, shortcut, or avoid.

- *Playfulness* is the idea that the patterned randomness of nature has a playful quality that we both need and desire in trails.
- *Harmony* occurs when natural shapes, anchors, safety, efficiency, playfulness, and physical factors all work together to sustainably support the desired trail experience.

Physical Forces

Human and natural forces caused by trail use and nature:
- *Compaction* is the downward force caused by trail use.
- *Displacement* is the sideways force caused by inevitable kicking, grinding, and acceleration forces of feet, hooves, and wheels.
- *Erosion* is the transport of tread* material by water or wind.

Tread Materials

How trail tread materials support trail use:
- *Tread texture* concerns the composition of soil, rock, and other tread materials and how that composition causes the tread to behave under physical forces in wet and dry trail conditions.

Tread Watershed

Unites all of the physical factors with trail and tread drainage:
- *Tread watershed* is both a description of a tread unit for drainage as well as a way to understand and relate twelve factors that determine tread drainage characteristics.

Each of these concepts is relatively simple in itself, making each easy to learn. The real power, however, is in how:

- They relate to and affect each other.
- They cover and relate both human and physical aspects in a single, integrated system.
- They create understanding of what works and what doesn't based on the "why" of their subjects instead of merely "how" or "what."
- They view complex situations as instances of their simpler underlying forces and relationships.
- They work with unique or previously unknown situations by examining their known underlying forces.
- They create a simple language that facilitates communication.
- They are logical and relatively objective, even with factors that are difficult to quantify.

The next five chapters discuss the concepts in each of the five categories in detail.

---

* In case you missed the definition of tread on page 9, "tread" is the actual surface on which we actively place feet, hooves, wheels, etc. In this book, "tread" is used to describe the physical travel surface while "trail" refers to the broad concept of a trail as a continuous travelway.

● **CHAPTER 3**

Human Perception

Basic shapes and relationships strongly influence our perception of trails and sites. These shapes and relationships are:
- Natural shapes
- Anchors
  – Edges
  – Gateways

Natural shapes, anchors, edges, and gateways describe the shapes of the world surrounding us and how we relate ourselves to our surroundings. These concepts also include how the described shapes and relationships make us feel.

Natural Shapes

Nature has a shape. It's very distinct. It's everywhere in the natural world, at all scales. It's instantly recognizable, yet unpredictable in its details.

Natural shape is this type of shape:

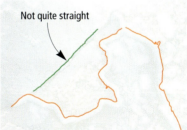

The knarled root (orange above) has an obvious natural shape. Yet the diagonal crack in the rock (green above) also has a natural shape since it is not quite straight. Natural shapes get their character by being unpredictable in the details.

13

A new named shape

You may be thinking that other words could be used to describe natural shape. Yet "curved", "curvilinear", "winding," and "irregular" are all inaccurate or imprecise. We need a new named shape to describe the precise nature of nature. "Natural shape" does this:

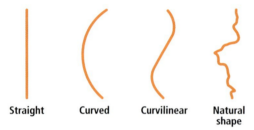

Straight   Curved   Curvilinear   Natural shape

In fact, natural shape is a "master shape" because it contains all straight segments, curves, and curvilinear shapes.

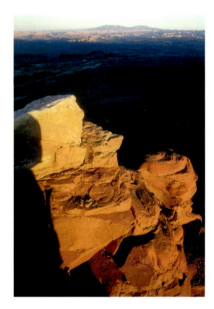

FAR LEFT *Natural shapes are clearly evident in the sun and shadow lines of the rock outcrop at left. The rock itself, the canyon background, and the horizon are shaped entirely of natural shapes.*

LEFT *Trees, mountains, river, and other vegetation are all shaped from natural shapes. When viewed closely, even the tree trunk isn't quite straight.*

And, of course, trails are most naturalistic when shaped with natural shape.

Natural shape exists at all scales. A pebble viewed closely looks like a boulder, or a mountain. To show natural shape at very different scales, the following four natural shapes are traced from prominent natural shapes in the photos/map below them. The dashed red outline in each photo highlights the natural shape that was traced:

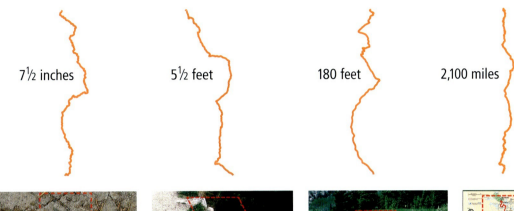

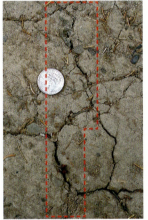

7½ inches of cracks in soil trail tread on the Split Rock Spur of the Superior Hiking Trail, Minnesota. An American quarter is shown for scale.

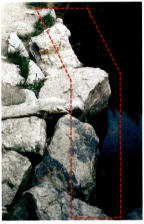

5½ feet of the outside edge of a stone-reinforced trail tread on the shore of Lake Haiyaha, Rocky Mountain National Park, Colorado.

180 feet of visitor-formed trail at Lower Cataract Lake, Arapaho National Forest, Colorado. The left edge of the trail is outlined.

The entire 2,100 mile length of the Appalachian Trail from Georgia to Maine.

Rough surfaces and rustic, primitive-style construction depend on natural shapes.

Natural shape also gives a definite shape to the physical aspects of words such as "rough," "rustic," "primitive," "rugged," "naturalistic," and "wild."

Because it is the shape of nature at all scales, natural shape is one of the most important and most often used concepts throughout this book and all three levels of this trail design system.

HUMAN PERCEPTION

Anchors

An anchor is any distinct vertical feature in the visible area. The more a vertical feature attracts and holds your attention, the stronger it is as an anchor. Similarly, the more that the trail or site reacts to it or wraps around it, the stronger it is as an anchor:

Because the trail wraps around it, the rock outcrop is a very strong anchor for this trail. Smaller rocks and trees on the right make additional anchors.

Anchors give a trail a visible reason to be "here" instead of "there." We tend to be drawn to anchors and feel comfortable in their presence—especially naturalistic anchors.

FAR LEFT *Anchors with natural shape feel most comfortable. In particular, anchors that flare at their bases—like trees—are more attractive and stable than anchors that go straight up, like walls of most buildings. Why? A flared base creates a natural feeling of stability as well as natural shape in the vertical dimension.*

LEFT *A lone tree is a natural attraction in this meadow. Because the trail changes shape to wrap around the tree (natural shape), the trail **incorporates** the tree. Incorporation greatly strengthens the tree as an anchor, highlighting it as a site feature and a natural feature to be appreciated.*

Our eyes are drawn to points of greatest contrast, so size, distance, and distinctness (contrast) matter. A distinct but distant anchor such as a lone tree in an open prairie is a strong anchor. If the trail comes anywhere near this tree, people will want to go to the tree.

Most anchors, however, are relatively ordinary site features. Even a small rock at the edge of the trail tread helps anchor the trail, especially if it is the only rock nearby.

Two special types of anchors affect us even more strongly: edges and gateways.

Edges

Edges are extended anchors. Common types of edges are the edge between meadow and forest, land and water, valley wall and valley floor, cliffs, and edges between distinctly different vegetation areas or ecosystems with a vertical element.

People love to be on edges, especially strong and distinct edges. Being on the edge gives you the experience of both worlds.

Not surprisingly, edges shaped with natural shapes—like this cliff edge—feel the most comfortable and anchored.

Being on an edge, we experience the realms on both sides. The sharper the contrast between the two sides, the more we usually enjoy being there.

Like anchors, edges can be natural or constructed. If constructed, edges with natural shape are more interesting. Constructed edges that intimately respond to their sites are most interesting of all.

This is all about edges and anchors. The water, the stone retaining walls, and the abrupt edge of the trees are all strong edges. Vegetation between the trail and this constructed pond helps anchor the trail while creating a comfortable buffer between trail and water.

Although current ecologic protection practices discourage trails this close to water, humans innately desire this experience and treasure the opportunities.

Distant yet visible geologic features such as valley walls and mountain ranges can serve as both edges and anchors for entire sites or even regions.

Mountains form a clear edge anchoring the trail and valley. The mountains even provide a visible goal that makes travel through this recent logging clearcut much more appealing.

The trail also forms its own edge in the site. The trail edge feels most anchored and integrated when formed with natural shape. Imagine how unintegrated and unanchored a straight trail would feel here.

This ATV trail in a savanna is enlivened purely by natural shapes, anchors, and edges. Its natural shape contours around slopes and roughly follows (is anchored by) the woodland edge. If the trail had gone straight through the center of the grassland, it would have far less intrinsic interest.

Crossing edges

Of course, we don't have to just follow edges. Some of the most engaging trails cross various edges in all kinds of ways.

For instance, the savanna trail above would be more interesting if it went into the trees occasionally. How long the trail parallels an edge, how it angles across the edge, and how it goes back and forth across an edge all affect how we perceive the site.

We can also align trails to shape sightlines with varying views relative to the edge. Interesting trails shape their rhythm and flow by playing with sightlines, frequently changing directions, and exposing changing views and viewsheds. This doesn't require a world-class site. With a little care and consideration, just crossing in and out of the trees can provide a lot of variety.

The next page compares the shapes and feelings of different ways of following, approaching, and crossing edges.

Edge crossings between bright sun and deep shade.

Following, Approaching, and Crossing Edges

Much of the feeling of a trail comes from how it relates to site edges. Each type of relationship has its own feel. The most engaging trails have sequences of many or all of these edge relationships:

Skirt edge in one place
Approaches and skirts the edge in only one place. This is often used to look into but not enter a sensitive area.

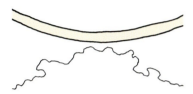

Follow edge without crossing
Respects the edge and is anchored by the edge. Prolongs the richness of being on the edge.

Cross edge head-on
Feels abrupt, maximizes feeling of sharp contrast.

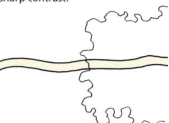

Cross edge obliquely
Much softer than a head-on crossing, feels relaxed and gentle. More naturalistic than a head-on crossing.

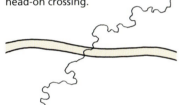

Follow edge on both sides of crossing
Lets us experience the edge from both sides. Most interesting when the trail can follow the edge for awhile on both sides.

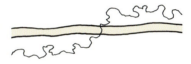

Cross edge repeatedly
Creates dynamic excitement and feeling of rapid change and progress, feels well integrated into the site.

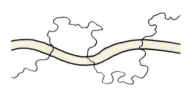

Head-on crossing: This trail intentionally crosses between grassland and woodland as abruptly as possible to emphasize the dramatic contrast between the two ecosystems.

Following edge on both sides of crossing: The trail follows and criss-crosses this stream through a narrow canyon. Here, the mortared wall (which keeps the stream in the channel during spring runoff) leads the eye directly to the well-anchored bridge (gateway, see next page). The plain bridge becomes more special through its context.

Gateways

Gateways are where the trail is clearly constrained on two or three sides: left, right, and/or above. The more the trail feels like it squeezes through, the stronger the gateway. Gateways are also anchors and may have qualities of an edge.

FAR LEFT *Boulders form a gateway. Passing through the sequoias also feels like passing through gateways—one of the pleasures of groves of tall trees.*

LEFT *A double gateway: first the two trees, then the railed bridge.*

FAR LEFT *Abrupt narrow, low-clearance passage from shady forest to open meadow forms a gateway constrained on all three sides.*

LEFT *Even the very average gateway of a tree arching over the trail creates a sense of passage.*

Gateways create a sense of passage and distance. In architecture, cathedrals and temples often use one or more gateways—low passages between larger, higher spaces—to help people spiritually "cleanse" themselves of the outside world. On trails, each gateway increases the feeling of distance traveled, making the trail seem longer than it is. A significant gateway near the parking area also makes it psychologically easier to adapt from the "car world" to the "trail world."

Gateways also intensify focus on the gateway itself, especially if visitors need to adjust their travel because of it. If the gateway is a natural site feature, it intensifies our recognition of the natural world.

The Power of Combinations

Where are the most stimulating places? Places with combinations of multiple natural shapes, anchors, edges, and/or gateways capture our attention. Harmonious presence of multiple scales, layers, materials, textures, colors, and overlaps also creates richness.

The power of combinations helps describe the appeal of places like Vernal Fall in Yosemite National Park, left. Here, combinations include dramatic edges of weathered stone, fast water, and steadfast trees; variety of materials in rocks, water, and trees; the canyon as a gateway view on the rugged (natural shaped, anchored) trail; layered foreground, middle ground, and background; wide variety of natural shapes, anchors, and edges at many scales; contrasts in color, light, and texture, including water and mist; and, of course, the thunder of crashing water.

Like poetry, every element in a combination is part of several instances of natural shapes, anchors, edges, gateways, scales, layers, materials, textures, colors, and overlaps. When all of these aspects harmoniously work together, the whole is much greater than the sum of its parts.

Most combinations are far more ordinary than Vernal Fall. Yet even humble combinations enliven trails wherever they occur:

FAR LEFT *Rocks form anchors, edges, and a gateway for the trail; water (to the left of the hiker) is another edge; the mountain background is both an edge and a strong regional anchor; the tall tree at right and smaller shrubs at left add additional anchors. The combination quietly enlivens the trail experience.*

LEFT *The boulder is both an anchor and one side of a gateway; limited sight distance creates layering and a sense of anticipation; other rocks and trees anchor the trail in horizontal space. Note that all of this was shaped, not "built," simply by choosing where the trail winds through the site.*

In addition, the best trail structures help form harmonious combinations:

Rough-hewn logs fitted to rocks make a well-anchored bridge full of natural shapes. Most of the corners and cut log ends are not quite square, relaxing the geometry. The dogleg adds natural shape to the entire bridge form. Plus the bridge itself is a gateway.

Stone steps wrap around existing rocks and shrubs. The steps themselves are anchored not only by site rocks and shrubs, but also by small plants at their bases. This staircase sees up to 1,000 trips per day year round and has been here for 15 years.

Plants improve combinations. Near a trail shelter on a popular trail, this rustic (natural shaped) stone wall protects an island of trees and plants. The stones anchor the trees and plants behind it, while the small plants in front help anchor the wall itself.

These stones not only stabilize trail tread at each end of the bridge, they also anchor and enliven an otherwise very plain bridge. Without the visible stone, the bridge would lack any anchored connection to its site. The stones add a third color, material, and texture to the wood deck and soil trail tread. Finally, the stones form a threshold we have to cross, accentuating the bridge as different from the regular tread and augmenting the gateway effect of the bridge.

Conclusion

Natural shapes, anchors, and the powerful anchoring effect of edges and gateways organize space. These specific shapes and relationships make us feel things as we see, cross, and pass them. They relate the trail to the site, anchor the trail in space, provide reasons to be here and not there, and create points of attraction. And having specific names for them makes it far easier to talk about trails and sites in simple, objective, spatial terms.

In the next chapter, we look more at how our feelings affect our trail experience.

CHAPTER 4
Human Feelings

Recreational trails are about our trail experience. In other words, our feelings. Yes, emotions—the amazing ability we all have to instantly feel a complex situation, and for different people to feel it in much the same way.

Yet we often find it hard to talk about feelings and experience, so we avoid it. It's especially hard in a government or agency setting where recreation—literally, "re-creation," a spiritual phenomenon—is supposed to be integrated with science or best management practices (BMPs) that generally avoid feelings.

We all know, however, that even governments base decisions on feelings. Remember "pursuit of happiness" and "our sacred honor" from America's Declaration of Independence? *Feelings motivate people like nothing else.* And recreational trail design is largely about feelings. We need to incorporate feelings into trail design in a bureaucratically palatable way.

Feelings vs. opinions

The key to appropriately using feelings in design is that different people *do* feel pretty much the same way about the same situation. We can trust that feeling and legitimately base decisions on it. Note, however, that we're using feelings, not *opinions*. It can be empirically shown that a group of people can have pretty much the same *feelings* about a situation, but a huge range of different *opinions* about the same situation.* In fact, by working with feelings instead of opinions, a group of strangers can often have 90-99% agreement.† This makes feelings an extremely powerful design tool.

In the following, remember that *feelings MUST NOT be confused with opinions.* Also, feelings should augment reason and logic, not replace them.

Bureaucratically palatable feelings in trail design

For working with feelings, natural shape, anchors, edges, gateways, and combinations provide a good start. They are all spatial relationships that cause us to feel certain things. Let's take that a bit farther.

For trails, we can represent most of our feelings with four bureaucratically palatable core concepts:

- Safety
- Efficiency
- Playfulness
- Harmony

* Christopher Alexander, *The Timeless Way of Building* (New York: Oxford University Press, 1979), pp. 290-92.

† Ibid, pp. 292-97.

These concepts relate to and balance each other:

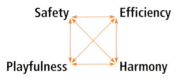

All four concepts are part of any situation. But first, let's look at each concept separately.

Safety

Our perception of safety depends on the individual person and the context. Obviously, a steep, narrow, rocky cliffside trail may feel safe enough for a rock climber but not for a family with young children looking for an easy trail, for a wheelchair user, or for someone with acrophobia. Similarly, a trail may feel safe enough when dry but not when it's wet, muddy, snowy, or dark. Unexpected hazards could also exist such as slope failure on a steep slope, falling trees, attack by wildlife, etc. Hence:

1. The feeling of safety is partly an individual's perception, and
2. Actual and perceived safety depends on all physical and perceived factors in the context.

Given the above, we can say our feeling of trail safety includes, but is not limited to:

- Being inside one's comfort zone—feeling physically safe or having risks within one's own expectations and limits
- Having a sufficiently stable and comfortable trail tread (stability, smoothness, traction) for our expectation and ability
- Having sufficient clearance around obstacles and edges
- Having a sufficiently comfortable buffer or protection from danger, or sufficiently low likelihood of danger (from wildlife, other people on or off the trail, steep dropoffs, site hazards, etc.)
- Having the trail as a continuous and sanctioned passage through a larger area (i.e., little or no fear of losing the trail or getting lost, having permission to be there)
- Being in one's comfort zone regarding other trail visitors (i.e., user conflicts, type and number of visitors)
- Expecting constructed features to be safer than natural features (although we expect some risk even from constructed features)

Safety and risk

While there's an essential need for safety, there's also a primal desire—even a psychological *need*—for risk. Risk requires judging conditions, assessing and testing one's abilities, and making personal decisions with one's personal safety at stake. Dealing with and performing in the face of manageable risk improves self-esteem, provides personal satisfaction, and creates a sense of accomplishment—all of which are essential for mental health and many of which are often sought from outdoor re-creation.

Whether or not you think this hiking-only trail is safe depends partly on you. From the tread wear pattern, though, we can see that no one seems to have walked on the outside edge of the tread.

Having a railing—even steel pipes—is welcome on this cliffhanger trail.

Depending on the context and visitors' expectations, we commonly can and do allow—or even seek to include—risky trail situations that fall within the above list. An appropriate level of risk is gratifying, whereas trails that don't leave us enough room to make personal decisions may disappoint or encourage visitors to find their own risks off the trail.*

* It's possible to use natural shape, anchors, edges, gateways, and combinations to form trails that cause more personal decision making and feeling of accomplishment with little or no increase in risk. See Book 2 of this series.

Two accessible trails with very different approaches to safety and risk:

FAR LEFT *Timber curbs keep wheelchair users on crushed stone tread on level valley bottom. Trail stays far from the river [visible in right background]. The trail feels stifling and frustrating, making everyone feel invalid regardless of their mobility. In a world-famous park, this trail is seldom used.*

LEFT *Trail specifically designed and built to be accessible to sport wheelchair users yet feel like an ordinary hiking trail. The stone wall is just a retaining wall, not a curb on this narrow, outsloped, native soil tread. The trail has no protections that a similar trail for able-bodied users wouldn't have.*

Efficiency

Visitors want to feel that the trail isn't wasting their time and effort. Efficiency concerns our willingness to stay on the trail vs. how much we might want to bypass or shortcut it. Efficiency includes the ideas that:

- Trail, tread, and tread structures need to be faster and easier to use than leaving the trail.
- Destination trails should be reasonably direct given site features and trail purpose.
- Visitors often want to move as fast as the trail, site, and their modality allow (i.e., speed limits don't work).
- "The trail to the restroom should always be straight."

Specific trail problems caused by visitors seeking higher efficiency include:

- Avoiding an excessively muddy, wet, rocky, or difficult main tread by traveling on the vegetated edge or forming parallel treads
- Bypassing tread structures such as waterbars and inconveniently spaced steps that are easier to bypass than use
- Shortcutting switchbacks—especially shallow switchbacks—that are faster to shortcut than use
- Speeding (for wheeled modalities) on trails with long sightlines, straight alignment, and wide, smooth treads—it's simply more efficient to move fast where one can
- Cutting corners and shortcutting broad, intervisible curves

If we don't want these things to happen, the best solution is to shape trails that don't set the stage for these behaviors.

Problems caused by lack of efficiency:

FAR LEFT *Timber risers are not spaced at pacing intervals, causing visitors to have to break pace to use them. Some find the risers too awkward and avoid them with a new tread at left. [Note: this trail was later completely rebuilt without risers.]*

LEFT *Visitors bypass deeply eroded tread.*

Problems caused by lack of efficiency:

FAR LEFT *This wide spot in an otherwise narrow, smooth tread resulted from visitors trying to bypass a rocky tread section. Some find it more efficient to travel on the undisturbed, smooth soil edge than to pick their way through the rocks.*

LEFT *Backcountry switchback with major efficiency flaws. It needs long constructed barriers because its upper leg is too close to level. With a sideslope this steep, few would shortcut the switchback if the upper leg had a steeper grade and it was faster to follow the trail than to shortcut it. Also note the sudden increase in tread grade halfway down the upper leg. All of its problems are design flaws.*

*Problem caused by **too much** efficiency:*

Long, straight, wide, smooth trail sections with wide clearance and long sightlines enable ATV users to reach high speeds. In braking hard for this curve, they drag tread soil, creating bumps in the tread much like large washboards. Holes are beginning to hold water and that will someday cause other bypass problems. The superelevated (banked) curve is also caused through displacement although this is not necessarily a problem (see page 41).

Playfulness

Playfulness helps balance efficiency and safety. It's the main difference between trails and roads, so it's one of the most important aspects of recreational trails.

In trails and sites, playfulness has natural shapes—crooked lines, uneven spacing, variety. Not everything is revealed at once. Things have depth, layers, ins and outs, roughness. The trail reacts to and wraps around site elements—acknowledges them, incorporates them—instead of railroading through as if they were hardly there. The tread is somewhat uneven and random. Physical aspects that are dead straight, dead square and dead level *feel dead*—never playful.

Playfulness includes:

- Anticipation, excitement, curiosity, surprise
- Peacefulness (especially after more energetic trail sections)
- Quirkiness
- High degree of reaction to the site—tight integration with the site
- Natural shapes, anchors, incorporated anchors, varied ways of following and crossing edges, gateways, and combinations
- Variety and contrasts in feelings as you proceed along the trail
- Desire for dramatic spaces and sensations (often through natural shapes, anchors, edges, gateways, and combinations)
- Contrasts between physical sensations (sight, sound, smell, touch, climbing, dropping, turning, etc.)
- Appropriate timing, rhythm, and flow between sensations—different sensations occurring at a comfortable rate, not too fast or too slow for expectations

Natural shapes, anchors, edges, gateways, and combinations are prime sources of bureaucratically palatable playfulness. As Book 2 in this series shows, this type of playfulness can actually **increase** sustainability.

Wrapping around site features, like the way the tread wraps around the rock and goes around the bend, is always playful. The more playful a trail can be, the more engaging it can be, and the more things we tend to notice, the longer it will seem and the longer it will take to travel it. Perhaps we don't need so many new miles of trails as much as we need more playful trails that seem longer than they are and take longer to travel.

A playful accessible trail: Rocks anchor the trail and the tread wraps around and responds to rocks, including one in the center of the trail [left foreground]. Wheelchair users can leave the tread if they want. Tread width varies. The lake in the background anchors the far end. Designed by the author, the site is alpine tundra at 12,840' elevation (Summit Lake on Mt. Evans, Colorado).

Even without strong anchors, trails gain playfulness by incorporating natural shapes in both the horizontal and vertical. This segment could have been straight, but the horizontal and vertical roll makes it more fun and draws the visitor's attention to plants and site aspects that wouldn't be a focus along a straight trail. The dip in this roll also defines a tread watershed (see Chapter 7).

Harmony

Harmony is our feeling of overall appropriateness—of the trail being comfortable in its site, of the trail being comfortable for us, of us not wanting to change anything. It's highly dependent on the context and, like all of our feelings, we feel it in degrees. Instantly.

We feel the most harmony when everything—the site, tread, trail structures, trail usage, how we move—works together to support our desired trail experience. This includes our feelings of safety, efficiency, and playfulness as previously described.

At best, harmony increases our sense of appreciation, respect, and stewardship for the trail and site because we realize we're experiencing something special that uniquely exists in that particular context. High harmony creates enough energy to affect our behavior.

On the other hand, anything that doesn't support our desired experience—as well as anything that degrades the trail or site in the short term, such as avoidable erosion, visitor-caused damage, and rapid change—reduces harmony. We often feel these as "something wrong" or feel we want things to be different. Low harmony tends to invite abuse, degradation, and rapid change.

Harmony has, but is not limited to, the following characteristics:

- **Site integration:** The trail seems part of the site rather than just sitting on it. It respects, echos, reacts to, and responds to the site and trail use.
- **Mechanical integration:** The shape, mechanics, and function of the trail feel integrated with the site—not alien to it—and contribute to its playfulness as well as its safety and efficiency.
- **Support of movement:** Trail and tread support our movement and our modality with the desired speed and challenge (safety, risk, efficiency, playfulness, natural shapes).
- **Trail width:** Width is not too wide or too narrow for the trail purpose and desired experience. *Appropriate width is essential.*
- **Trail location:** Location and alignment of the trail is in harmony with the desired trail experience. Site and alignment supports the trail and trail use.
- **Stability:** Trail tread is stable with no rapid or unexpected changes such as from fast erosion, although slow change is acceptable as long as it can be sustainably accommodated.
- **Weathering:** The trail and its structures have blended with, reacted to, and been changed by the site—weathering, vegetation growth, etc. New trails cannot have the harmony that established ones can (part of the appeal of established and historic trails).
- **Peacefulness:** Change happens only slowly as the trail and site accommodate and adapt to natural forces and trail use. (Rapid change, usually disruptive, is the opposite of peacefulness.)
- **Rhythm and flow:** Spacing between attractive sensations and experiences is satisfying in both time and distance. Note that these attractions don't have to be major landmarks—most of them can often be formed all along the trail from natural shapes, anchors, edges, gateways, and combinations.

*Harmony is **NOT** about being smooth, new, or pretty. It's about being comfortable and stable because physical and psychological forces support the interaction between trail, site, and usage in its present state. This visitor-formed trail roughly following an old prairie fence (edge) has good harmony between trail, site, and usage. At its current type and level of use, physical and psychological forces are in equilibrium—hence the trail is stable, peaceful. In the context of this trail and site, the naturally shaped, rocky tread is a major contributor to harmony.*

The most harmonious trails let the site do all the work. Sure, a great site helps. Yet no matter how plain the site, the trail should always seem to be part of the site rather than seem imposed on it.

Lack of harmony is easily seen as "leakage" of unresolved forces. Here we see timber risers and obviously eroded tread as erosion control attempts—where erosion itself is a leakage of unresolved forces. The waterbar in the foreground appears to contain erosion to within this short section.

The logs in the grass to the left attempt to plug another leakage: visitors trying to avoid mud by traveling on the upper vegetated edge. When a new tread started to form, the land manager tried to plug the leak with logs without addressing the mud problem that caused it.

There is some harmony here, namely in the larger rocks that anchor the trail and in the overall feel of the trail in this site. The underlying problem, however, is that the trail is climbing faster than it should and captures more water than soils, grades, and use can sustainably handle (see Chapter 7). Partial, unintegrated, graceless fixes call attention to themselves and detract from our enjoyment of the site.

- **Natural materials:** Natural materials echo the site whenever feasible. Manufactured materials are allowed to be visible only where necessary or where the site and visitor expectations support their use. Larger installations of manufactured materials are shaped with natural shapes as much as feasible.
- **"Shaped" vs. "engineered":** Harmonious trails and trail structures are often rough around the edges. They have natural shapes at every scale as well as naturally shaped anchors and edges. This shapes a relaxed, informal feel that echoes nature. Structures have weathered surfaces, irregular edges, non-square corners, custom fit to an irregular site, and a relaxed geometry that reflects nature and lets us feel comfortable. The most harmonious trails are often the most humble, most "shaped" and least "engineered."
- **Generation of appreciation and stewardship:** By feeling contextually special, harmonious trails tend to engender a sense of appreciation and stewardship for the trail and the site that supports it.

More about what harmony is—and isn't

- As used here, "harmony" includes visitors' direct feelings—what we directly experience while on the trail, what we directly sense—but *not* our *opinions* about "how things should be." Hence ecosystem management, such as managing trails in sensitive habitat for flora

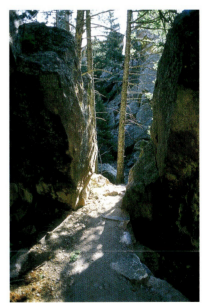

*Instant harmony: When you get to a tough spot, **take advantage of it.** This trail got out of a tight spot by climbing up to this natural slot and going through it. The result is an exceptional gateway and perfect harmony for the context of this steep, narrow hiking trail in rocky terrain.*

Lack of intimacy reduces harmony. While using old roads as trails can have land management advantages (see Book 3), trails on roads never have the intimate harmony or provide the satisfaction that a narrow, more naturally shaped tread could for any trail modality. Although they physically support us, roads don't cradle us. By being too wide, they make the site seem distant. By being too "built," they reduce our touch of natural shapes. Road-based natural surface trails also encourage faster travel and make the trail seem shorter than a narrower, naturally shaped trail would.

or fauna, is not directly considered under harmony as we use this concept in the Foundation Level. Why not? Because visitors could be impacting an endangered species, *but if we don't know it or feel it,* our impact doesn't reduce our sense of trail harmony which is formed by what's readily apparent to our senses. It may be decided to close the trail to reduce ecosystem impacts (an Upper Level decision) even though it has good Foundation Level harmony. (Ecosystem management and its effects on trail design and Foundation Level harmony are discussed in Book 3, *Managing Trails by Design: Integrating Stewardship, Sustainability and the Trail Experience.*)

- Although each trail should have as much harmony as it can, we expect more harmony from some trail contexts than others. Think of a national park trail versus a neighborhood trail. If our reason for being on a trail doesn't require a high degree of harmony, we happily accept lower harmony. And even a visitor-formed trail in a vacant lot can have a surprising amount of harmony. In fact, many of my own ideas, including some core ideas in this book, came from studying visitor-formed trails that have wonderful yet unplanned harmony.

- Harmony doesn't require perfection or a static state of no change. As stated above, part of harmony is that things *do* change, although slowly and gracefully in direct response to nature. Materials weather, natural surface trail treads constantly change (see the next chapter), and weather and water change contextual harmony. Nearly all trails require occasional maintenance (reshaping). And even a trail that relies on constructed drainage dips and waterbars for drainage can be in harmony if those structures are sustainable (Book 2). It's much easier to achieve harmony, though, if the trail can be stable through more naturalistic means (again, Book 2).

- Complete trail harmony is rather rare. Yet many trails can get fairly close. They feel comfortable—in tune with us and the site. They have the raw beauty of form following function while using and

Harmony depends partly on the visitor's modality. For the off-highway motorcyclists who formed this trail, the anchored and banked curves are completely in harmony with their higher speeds (see page 41). That this has been a motorcycle trail for 25 years without major maintenance is further proof of harmony for motorcycle use. The same tread, however, would not feel harmonious for pedestrians or equestrians, and is too narrow for ATVs.

Yet tread shape is only part of harmony, and this trail segment has a lot of harmony through natural shapes, anchors, and the generally comfortable feel of the trail in this site. Visitors using any modality could find harmony here if the tread were reshaped to suit their needs and expectations.

How Our Modality Affects Harmony

Recreational trails give us ways to engage our senses, including our sense of movement and how we move. Hence how we move greatly affects how much harmony we feel:

Time, distance and speed matter. If we're moving slowly, we desire places of higher harmony to be relatively close together so that we can experience more harmony in a given time. If we're moving faster, though, trails can have relatively less harmony per unit of distance because we can cross distance faster. In practical terms, faster-moving visitors can more readily enjoy trails and sites with lower harmony than can slower-moving visitors. High-harmony trails are most appreciated by slower-moving visitors. Simply stated, we seek a certain amount of harmony per unit of time.

Concentrating on movement reduces our need for harmony. While we're always aware of what's around us, the more that we're concentrating on our boots, horse, bike, or OHV, the less we feel a strong desire for high harmony. Hurrying or speeding along the trail can also reduce our need for harmony. Of course, even while concentrating on our modality, we still enjoy being in contexts with high harmony.

Tread shape and tracks are important to harmony. Each modality tends to shape trail tread to conform to itself but not others. For instance, pedestrians, mountain bikes, and ATVs each shape the tread in distinctly different ways (Chapter 5) and leave very different tracks. Efficiency and safety may also be involved. Naturally, we tend to feel more harmony from tread shape and width that cradles our modality than we feel from other tread shapes and excessively wide or narrow widths. The popular request for singletrack trails rather than old roads is a frequently heard example.

Our sense of harmony largely depends on our modality. From the above, visitors using different modalities will have different senses of harmony on the same trail. On an ATV, cruising fast down a straight old road, then climbing over a series of short hills and splashing across an unbridged stream can be a lot of fun, yet few hikers would enjoy the same trail as much.

Visitor conflicts often involve the above four points. Sharing a trail with others moving at different speeds, concentrating on movement at different intensities, having tread shaped in ways that don't suit our modality, and deriving enjoyment from different aspects than others can reduce our own sense of harmony. The Foundation Level helps us understand these conflicts and, to the limited degree possible at this level, suggests some ways to reduce them. See Book 3 for more on visitor conflicts.

incorporating nature and the site. They require little or no tread maintenance. The mechanics of the trail are thoroughly integrated with the site and contribute to its playfulness as well as its safety and efficiency. We can feel very comfortable just being on the trail, not even moving—just enjoying it.

Human Perception, Feelings, and Stewardship

Feelings drive a major part of trail design. Satisfied feelings of safety, efficiency, playfulness, and harmony lead to good trail experiences that help enhance our appreciation and respect for trails, sites, and natural resources in general. Appreciation and respect, in turn, help engender resource stewardship. And, of course, our feelings are based largely on how we perceive the site and how well the trail weaves itself into the site by using natural shapes and anchors.

On the other hand, unsatisfied feelings may leak out and lead to myriad trail and site problems. Feelings, unfortunately, are often omitted from land and ecosystem management processes, causing or exacerbating trail use problems (see Book 3).

As a concrete example of how trails can use perception and feelings to improve the trail experience and encourage people to stay on the trail, the next two pages compare the pros and cons of eight traditionally problematic trail features: switchbacks.

EXAMPLE

How Perception and Feelings Affect Switchbacks

So far, we've seen six of the eleven concepts of the Foundation Level. Let's use them to help explain and improve the design of one of the most problematic trail features—switchbacks.

Natural shapes • Anchors (plus Edges and Gateways) • Safety • Efficiency • Playfulness • Harmony

This switchback has no anchor to hold down the corner, has no strong natural shapes, has no noteworthy edge to separate upper and lower legs, and has little of a unique view to attract us to the outside of the turn when heading downhill. Visitors are slowly cutting the inside corner. Safety and efficiency are okay, but there's no playfulness and no great harmony. Hence it works functionally but doesn't add much to our trail experience.

We're attracted to the turn by the reward of a sweeping vista. Stone retaining walls below both legs form very strong anchors and edges that both support the trail and remove any temptation to shortcut. Safety and efficiency are both good on this hiking and horse trail. Contrasts in color, texture, materials, natural shapes, combinations, and views all add playfulness. Harmony comes from all of the above plus the fact that the implementation of the switchback is in harmony with the steep site—it makes wonderful use of the site. One of the best switchbacks I've ever seen.

This excellent switchback has strong natural anchors on both the inside and outside of the turn, creating a gateway. The two trees between the legs greatly help anchor the turn. The turn is wide on this mountain bike, hiking, and horse trail; and the approach rewards us with a vista. Trees and rocks on the outside of the turn help downhill bicyclists see the turn in advance so they can slow down ahead of time (safety). Both legs have relatively steep grades to help remove temptation to shortcut.

Stacked switchbacks can invite shortcutting by appearing inefficient, but these are much easier to use than to shortcut on this hiking-only trail. Boulders and rock outcrops on both sides provide excellent anchors and edges, and the switchbacks playfully bounce between them. Construction is only about two years old—the trail crew carefully preserved the green shrub between the uppermost legs (shrubs grow slowly in the dry climate here). The exit at top right passes through a natural slot (gateway) between boulders (shown on page 29). Excellent harmony results from use of natural site features and native stone, including stonework that is neither too formal nor too crude for this context.

This plain switchback could have been located anywhere on this brushy slope, but I intentionally wrapped it around one of the largest and tallest trees. The tall tree provides a goal to move toward—reaching it feels like a small reward. As you've noticed, the best switchbacks emphasize rewarding visitors who follow the trail more than punishing them by making shortcutting difficult. Safety, efficiency, playfulness, and harmony are all quietly evident.

*Switchbacks in open sites desperately need anchors and a visual goal to move toward. Without the boulder gateway in the turn, visitors would have little incentive to follow this switchback in a meadow. Yet the trail has been here for many years with no sign of shortcutting—a prime example of the power and value of anchoring switchbacks. It's always more harmonious when trails are designed such that visitors **want** to "do the right thing."*

*This may be the worst switchback I've seen. The lower leg is nearly level while the upper climbs too fast in sandy soil. Visitors going uphill don't even see the turn and try to continue into the trees, partly because the log step/retaining wall leads our eye away from the upper leg. Safety lacks when descending because we have to make a downhill step, sudden stop, and hard turn at the bottom of the steep upper leg. Efficiency lacks because we don't see the turn when climbing—there are no tread shape cues to indicate that there even **is** a turn—and we also want to shortcut when descending. There's nothing playful about how the switchback relates to the site or about how its parts are shaped. The carelessly formed stone retaining wall below the lower leg reduces harmony by not having any respect for the site, and the dead branches piled between the legs are an ugly attempt to prevent shortcutting. This switchback only degrades its site.*

On accessible switchbacks, closely spaced legs with shallow grades often feel inefficient, especially to able-bodied visitors. If a switchback is unavoidable, the best approach is to make the corner interesting so that we're rewarded by following the trail. In this old-growth forest in Oregon, there's no view and nothing unique about this particular spot. Because of the shallow grades, trying to wrap the switchback around a tree or shrub also wouldn't work—although there could have been a large anchor tree on the outside corner of the turn. But there is some native stone on site [far right], and if it were used to form a naturally shaped wall between the legs instead of the straight log, it would increase the appeal and harmony of this newly built switchback.

*The world is not to be put in order, the world is order incarnate.
It is for us to put ourselves in unison with this order.*
 – Henry Miller

God dwells in the details.
 – Mies van der Rohe, architect

● CHAPTER 5

Physical Forces

Several human-caused and natural forces act on natural surface treads. For our purposes, we can work with them as three forces:

- Compaction
- Displacement
- Erosion

While erosion—a natural force—tends to get the most attention, the human-caused forces of compaction and displacement play key roles in trail design and sustainability.

Compaction

Compaction is the downward force of visitors' weight, plus the weight of our modalities, on the tread surface. Hiking boot lugs, horseshoes, and knobby tires all concentrate weight in relatively small areas, intensifying compaction force.

Continual, cumulative compaction from year-in, year-out trail use is akin to beating the tread with a hammer—hard. Tread hardens and sinks relative to adjacent untraveled areas.

This trail tread in sandy soil shows effects of all three physical forces: compaction, displacement, and erosion.

Facts about trail compaction:

- Visitor-formed trails initially form through compaction:

Stage 1: Trampled vegetation begins to be stressed through physical impact and soil compaction.

Stage 2: Vegetation begins to die from abrasion and soil compaction around roots.

Stage 3: Tread vegetation and roots have died. Tread sinks from compaction and displacement.

- Nearly all trail treads sink from compaction—including constructed treads—although a number of factors determine how far a given tread sinks.
- Portions of the tread that frequently bear weight become more compacted than lesser-used portions. Typically, the tread center compacts more than the edges:

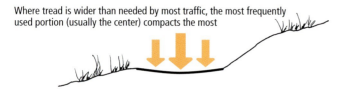
Where tread is wider than needed by most traffic, the most frequently used portion (usually the center) compacts the most

- For drainage, trail treads are often formed with a technique called outslope (see below and Book 2) that pitches the tread surface downhill while traversing a slope. This is similar to the pitch on paved roads that drains water to the sides. Soil and loose materials, however, are not as firm as asphalt or concrete, and compacting the tread center tends to cause outslope to fail. Outslope failure tends to occur fastest on newly constructed, highly compactible treads:

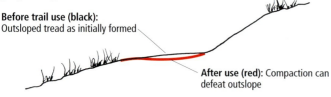
Before trail use (black): Outsloped tread as initially formed
After use (red): Compaction can defeat outslope

- Compaction hardens the tread by pressing tread particles into tighter contact and firmer bonding (see Chapter 6), reducing air and water spaces. Hence compaction makes the tread more resistant to displacement, erosion, and mud—a major benefit.
- Hardening the tread, though, also causes it to become less absorbent. Instead of being absorbed, nearly all water landing on the tread tends to pond and flow along it, increasing erosion potential:

BOTH PHOTOS Both compacted treads were photographed during light to moderate rain. All or most of the water on the compacted tread is from rain falling directly on the tread—not site runoff. Note ponding in low spots on both treads even from minor rain.

FAR LEFT This tread traverses a sideslope but compaction deepened the tread enough to pond water. Downhill is to the right.

LEFT With some tread grade, water flows downhill along the compacted tread in a site with little sideslope. The sunken tread is caused by years of compaction, displacement, and erosion.

Displacement

Displacement is human-caused horizontal movement of trail tread material (soil, gravel, stones, etc.) Friction of moving boots, hooves, and tires against trail tread unavoidably dislodges soil particles, gravel, and even rocks.

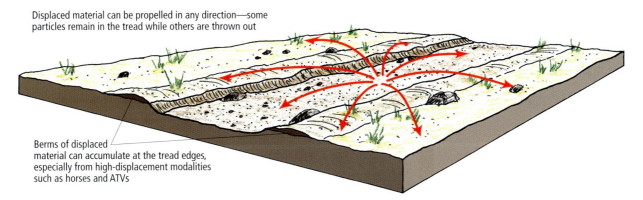

Displaced material can be propelled in any direction—some particles remain in the tread while others are thrown out

Berms of displaced material can accumulate at the tread edges, especially from high-displacement modalities such as horses and ATVs

ABOVE Displacement in a newly constructed mountain bike tread. The stick [center] spanning the sunken tread indicates the level of the original outsloped tread. A berm has formed on both sides of the rut. This tread displaced before it compacted.

ABOVE LEFT Displacement in progress: Along with a cloud of dust, these horses dislodge soil particles and small stones, kicking them around—and sometimes completely out of—the tread.

Displacement occurs when particles are propelled directly by the movement of visitors:

- Hikers inadvertently kick soil and rocks along and out of the tread.
- Horses flip their hooves and launch particles backward and to the sides, sometimes out of the tread. In loose or sandy soil, horse trails often develop distinct berms on both sides.
- Bicycles and OHVs flip particles in all directions. Faster speeds, quick braking, and quick acceleration exponentially increase displacement and the distance particles are thrown. On curves, most particles go to the outside of the curve (centrifugal force), tending to form superelevated (banked) curves.
- On muddy and loose-surfaced treads, displacement also includes leaving tracks and reshaping the tread by exerting more shear forces than the tread can resist.

Displacement invariably deepens the tread over time. While it can be reduced, displacement cannot be prevented with any natural surface tread unless the surface is somehow firmly bonded.

How Compaction and Displacement Interact

On a trail, vertical compaction force and horizontal displacement force always occur together and interact in specific ways:

1. Over time, compaction and displacement tend to modify *all* tread shapes. Many people naively believe that once a tread has been shaped, it will keep that shape indefinitely. A similar erroneous belief is that named tread shapes—outslope, crown, inslope—are somehow protected. Not usually. If all factors are favorable, treads can retain their original shape, but this is relatively rare. Over several years, you should expect a tendency toward the following changes in tread shape:

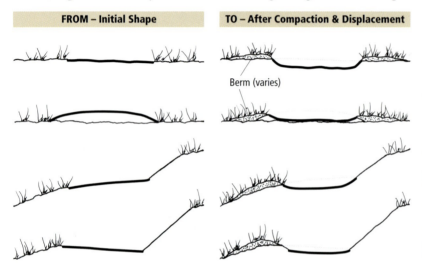

Ground level tread (no specific shape)
Tread deepened by compaction and displacement, berms on sides (varies). Note berm vegetation that tends to solidify and hide the berm.

Crowning
Crown sinks. Displaced crown material is thrown to the sides, possibly accumulating in berms.

Outslope
Outslope is defeated partly by compaction but mostly by displacement. Displaced material may pile in berm on outside edge. Restoring outslope requires extensive vegetated berm removal.

Inslope
Tread tends to level out as some displaced material fills in the low point against the backslope (here, right side). Some displaced material forms berm on outside edge.

2. Compaction hardens the tread, reducing displacement. Soils with high clay content or other firm bonding can become highly resistant to displacement. Hence compaction is desirable.

3. Compaction is limited but displacement is not. A tread will only compact and sink so much before further compaction force creates no further tread shape change. Tread shape change from displacement, however, can continue for as long as the trail is in use, constantly grinding and sinking the tread. Hence displacement can cause more long-term tread shape change than compaction.

Eventually, structurally useful tread materials resist further compaction…

…but have no resistance to grinding displacement except their own hardness/surface bonding. So displacement can potentially continue forever.

4. Tread material characteristics are major factors in compaction and displacement. As discussed in Chapter 6, some tread materials resist compaction and/or displacement better than others. *Hence a tread with one type of material can behave very differently from a tread with another material even when other factors are similar.* Tread material characteristics must be considered in all natural surface trail design (Chapter 6).

5. On grades, displacement increases and compaction decreases. The steeper the grade, the more that downward force is deflected into displacement force grinding on the tread. Assuming that the tread is firm and stable:

Level Tread
Compaction force goes straight down with gravity. Also see #7 below.

Moderate Tread Grade
Gravity causes some compaction force to "slide" downhill, adding to displacement force while reducing compaction force.

Steep Tread Grade
Much of the compaction force is deflected into downhill displacement force, leaving relatively little for compaction.

Hence the steeper the grade, the more displacement will occur and the less the tread will be hardened by compaction.

6. On grades, displaced particles creep down the tread. Even without erosion, gravity accelerates small rocks and soil particles headed downhill while it decelerates particles heading uphill. Hence displaced particles tend to creep downhill with each displacement. The steeper the grade, the faster they move downhill. See photo at right.

7. In muddy or loose tread materials, compaction force can displace the tread. We've been discussing longer-range, long term displacement that excavates the tread with use. Yet muddy tracks and tread shape change in soft or loose treads (mud, pure sand, round gravel) is also displacement. Here are the mechanics of what we see in the field:

On this very steep trail, small displaced rocks readily roll downhill. Much of the material clogging this waterbar rolled down the trail from displacement. This tread is only about a year old but will become a major problem in a few years. Also see "Dip Sustainability" on page 60.

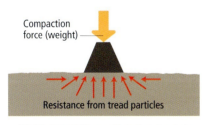
Stable Tread (firm soil and rock, etc.)
Friction and bonding between tread particles is enough to resist compaction force. In essence, tread material "pushes back" from all sides with enough force to resist compaction force.

Unstable Tread (mud, round particles)
Friction and bonding between tread particles isn't enough to withstand compaction force. This results in shear forces that displace the tread, squeezing material upward while allowing the weight to sink into the tread.

8. Fast-moving wheeled modalities tend to superelevate curves. By centrifugal force, bicycles and OHVs moving through curves tend to displace material to the outside of the curve. Unless the curve is already banked, material displaced from the inside of the curve tends to land on the outside, forming a banked curve over time. "Superelevation" is the engineering term for tread that pitches down

An extreme case of displacement on a muddy, unsanctioned trail. Fresh ATV tracks are from the ATV I was riding.

Mountain bike: Superelevation forming by use on an outside curve in a new, originally outsloped mountain bike trail.

Motorcycle: Superelevation forming by use on a visitor-formed motorcycle trail. Soil on the inside curve displaced to the outside, exposing tan-colored subsoil.

ATV: Superelevation forming by use on a new ATV trail only a few months old. Tread will keep displacing to the outside until the entire curve is superelevated by use.

toward the inside of curves. Tight curves, fast travel, loose tread material, and high traffic all tend to increase visitor-caused superelevation.

Once formed, though, superelevated curves can be very stable against further compaction and displacement. If the superelevation pitch is appropriate for visitor speed, centrifugal force adds to compaction force (beneficial). Displacement is greatly reduced. Also, centrifugal force tends to cause any launched particles to quickly fall back into the banked tread. Hence superelevation uses physics to create a self-sustaining shape for curves with wheeled modalities.

To some, the concept of superelevated curves on trails evokes images of racetracks, excess speeds, inappropriate use, and management problems. *Yet superelevation tends to happen whether you want it or not.* If you understand it and harmoniously incorporate it into trail design and shaping—either build it intentionally or design the trail to sustainably develop it—you can *use* superelevation to help make the tread more stable and improve the visitor's experience (by increasing modality-based harmony) at the same time.

Physics used—not fought. This intentionally superelevated switchback for mountain bikes increases tread stability, rider safety, and rider enjoyment all at the same time.

A short but sharp curve in a former road was successfully superelevated to stabilize it against ATV and motorcycle displacement.

Compaction & Displacement by Modality

Each modality exerts compaction and displacement forces differently. We can rate the compaction and displacement forces exerted by each modality and compare the tread shapes each tends to form over time. Why is this important? Because how each modality exerts compaction and displacement helps determine what's necessary and possible in the Middle and Upper Levels as well as how other Foundation Level factors will affect the trail.

Wheelchair
Compaction: MEDIUM
Displacement: VERY LOW

Medium compaction from narrow tires bearing weight of person + chair. Tread should be strong and firm enough to support the tires without displacement. Most use is often by human feet.

Human Feet
Compaction: LOW-MEDIUM
Displacement: LOW

Soles that are essentially flat side-to-side form a flat-bottomed tread. Low displacement from slow speed and rolling motion of walking. Compaction is concentrated toward the center.

Horse and Stock
Compaction: VERY HIGH
Displacement: MEDIUM-HIGH

Flat-bottomed, hard-edged hooves make a highly compacted, hardened, flat tread. Tread can be very narrow (14") as horses follow each other. Flipping hooves can cause high displacement with berms next to the tread.

Mountain Bike
Compaction: MEDIUM
Displacement: MEDIUM

Riders wander side to side but concentrate compaction and displacement toward the center. Tread becomes a gently rounded swale with displaced material adjacent and nearby. More displacement in looser soils.

Off-Highway Motorcycle
Compaction: MEDIUM
Displacement: EXTREME

Riders with 4-5"-wide tires concentrate compaction and displacement toward the center of the tread and sometimes in a deeper wheel rut **(dashed red)**. Knobby tires grab and dig in. Generally fast speeds with quick acceleration and braking can cause extreme displacement.

All-Terrain Vehicle (ATV)
Compaction: MEDIUM
Displacement: HIGH-EXTREME

Wide but small diameter, high flotation tires spreads vehicle weight, reducing compaction. High power, quick acceleration, and braking with moderate to fast speeds can cause high to extreme displacement.

Off-Road Truck/SUV (ORV)
Compaction: VERY HIGH
Displacement: LOW-VERY HIGH

Wide tires and very high compaction from weight of trucks smoothens, hardens, and levels the tread. Slow speeds cause little displacement on compacted tread, but higher speeds or wheel spinning can cause high displacement.

Hiking tread: *Compaction (mostly) and displacement of the top layer of soil exposes rocks. Note sunken tread with a relatively flat bottom.*

Horse tread: *Sandy soil displaces easily under flipping hooves, forming berms on both sides. New vegetation grows on the berms.*

Mountain bike tread: *Tread is forming the typical gentle swale of bike trails. Curves such as in foreground are becoming banked from centrifugal displacement.*

New ATV tread: *Roots are exposed from compaction and displacement. Note displaced loose soil at tread edges and in the center.*

Erosion

We all know what erosion is—water or sometimes wind moving with enough force to transport tread particles. Yet essential aspects of it are not widely understood:

1. We *accommodate* erosion rather than "fight" or "control" it. Language shapes our thoughts and attitudes, so it's critical to use appropriate words. "Fighting" erosion, however, is inappropriate. Why? Because erosion is a natural force, as natural as the Grand Canyon, the Mississippi River, Niagara Falls, and your favorite creek. Hence fighting erosion is fighting nature. "Controlling" erosion isn't accurate, either, since we can't, for instance, control how much it rains or control exactly how much erosion we allow before we allow no more.

LEFT *The Yampa River in Dinosaur National Monument, Colorado—one of countless examples of natural erosion shaping the surface of Earth.*

Instead, we *accommodate* erosion. We expect some erosion to occur on trails because we can't prevent it, as discussed below. But we shape trail, tread, and site contexts to help limit how much erosion can occur. Ideally, this shaping is as naturalistic (and low maintenance) as feasible and takes human perception and feelings into account. Think of this in terms of harmony as described in the previous chapter—using nature to help achieve harmony, including accommodating erosion. We'll discuss this more throughout the rest of this book and Book 2.

2. Small erosion events nibble away at the tread. An erosion event refers to the amount of erosion occurring during a one runoff event.

Most erosion events are small. They're caused by relatively small or moderate amounts of water and cause little visible erosion. Yet even small events carry vegetative and forest litter and dust down the tread, often depositing it in low points. Over years and decades, these tiny events add up to distinct tread loss.

#2. Even a light rain causes erosion. Over time, numerous light to moderate rains can erode significant amounts of tread.

3. Tread erosion starts sooner than you might think. Repeated empirical observations (standing out in the rain) show me that small erosion events can erode tread as little as 7 running feet below a local high point (crest) in the tread such as a drainage dip, waterbar, or natural crest. See photo at right. By 15 feet along the tread, even low flows on gentle tread grades of 4-5% carry dust, sand, and other particles.

4. Small erosion events tend to clog narrow-outlet trail drainages. The small flows of dust and vegetative litter tend to collect at points of concentration. These points include the narrow outlets of drainage dips and waterbars, especially when there is enough water to deliver debris but not enough to flush it through. In addition, water tends to cement the dust and debris together into a plug that solidifies before the next erosion event. Every time this repeats, the plug gets larger. (Also see "Dip sustainability" on page 60). Hence *a sequence of small*

#3. During light rain, visible erosion of vegetative litter begins only 7 feet below a local crest—even on a gentle grade with a compacted tread.

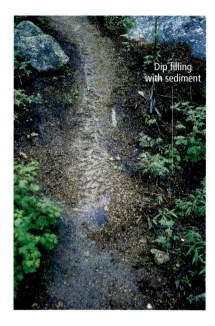

FAR LEFT #4. Small flows gradually fill a drainage dip [lower right] with vegetative debris and small tread particles. Ripple marks in debris left by a small flow are still visible [center]. Small flows have enough power to flow particles into low spots but not enough to flush them through.

LEFT #5. Where the site, trail, and tread context didn't limit it, intense rain caused severe tread erosion. It was predictable—just a matter of time.

flows can cause failure of narrow-outlet tread drainage devices even without major erosion events.

5. Large erosion events can scour the tread and tread drainages with narrow outlets. Occasionally, large, "100-year," and "500-year" events occur. Torrents of rain from a stalled thunderstorm or hurricane pour from the sky, run off the land, fall onto the trail, and erode it, scouring out large amounts of tread material. Heavy flows also tend to scour and/or overwhelm narrow-outlet tread drainages that successfully handled smaller events. Statistically, it's only a matter of time before one of these events happens in any given location, and global warming is expected to make major events more likely.*

How Erosion, Compaction, and Displacement Interact

1. Compaction and displacement set the stage for erosion. Combined with the removal of vegetation and roots, compaction and erosion tend to cause tread to develop a rut that can catch and carry water. Unless a number of factors are favorable (see the next two chapters), it's difficult or unfeasible to prevent this tread rutting. Hence it's prudent to plan to accommodate erosion caused by water running for limited distances along the tread. This is introduced in Chapter 7 and discussed in detail in Book 2.

* "'We see an increase in overall precipitation of 5 to 10 percent over the past century, but the increase is especially prominent since the 1970s,' said Tom Karl, director of the National Climatic Data Center in Asheville, N.C.

"'But then if you look more closely at how precipitation is coming about, what we're seeing is that the increase is coming primarily though an increase in frequency and intensity of heavy and very heavy precipitation events,' Karl said."
– Joan Lowy, "Global Warming Is Seen Around the World", *(Boulder, Colorado) Daily Camera*, 9 May 2004, sec. B, p. 7.

LEFT Only nine years old in this photo, this eroded tread was originally built as new outsloped tread. Compaction and displacement, primarily from mountain biking and hiking, quickly defeated the outslope in sandy loam soil. The depressed tread caught and carried runoff water from large areas of the slope, resulting in deep erosion.

2. Compacted tread helps minimize tread erosion. The hardening of compaction helps harden a tread against erosion. Hence compaction helps reduce both erosion and displacement.

3. Displacement tends to increase erosion. Erosion takes the loose particles first, so displaced particles are the first to go.

4. Compaction and displacement can hide evidence of erosion. Fresh tread erosion is easy to spot. After a few visitor trips, however, feet, hooves, and wheels grind off the hard edges and fill in the ruts, smoothing the tread and hiding erosion evidence. After a while, all that's evident is that the tread is lower than before—but with no "smoking gun" showing erosion as the cause. Unless you know what to look for, you might not even notice that the tread *is* eroding.

5. Compaction and displacement can hide evidence of deposition from erosion. Eroded material from upper segments of a tread can be deposited on the tread itself at the bottom of the grade. (See photo at lower right.) When this happens, the compacted and displaced tread rut at the bottom acts like an empty lake into which eroded material flows. The debris-filled tread then becomes level with its sides or with the berm on one or both sides. Once visitors compact and displace the filled tread, it can seem as if the tread has been that way all along. You, however, can spot deposition by looking for low spots where the tread surface doesn't appear sunken by compaction and erosion.

Tread buildup by deposition occurs in drainage dips, waterbars, drainage crossings (for typically dry crossings), and in longer tread dips or near-level tread segments below steeper segments.

The built-up tread itself, though, tends to rut through compaction and displacement, leaving a new depression to fill with the next erosion event. If the process keeps repeating, the filled tread can become higher with each filling. Eventually a raised causeway with vegetated sides can form—vegetation growth is almost inevitable since it's well watered.

Physical Forces in Context

Compaction, displacement, and erosion are the three physical forces that affect trail treads. So far, we've looked at the nature of each force along with its basic causes, effects, and tendencies. While each force is simple, each has multiple factors and interactions with other forces and trail/site aspects.

From the photos and graphics in this chapter, it may seem that natural surface trails are doomed. Frankly, some are. Yet remember that we're looking at tendencies. How much those tendencies actually occur in any given context depends on the context.

In the next two chapters, we'll explore more of the context in which these forces occur on trails: soils and tread materials (Chapter 6) and tread watershed factors (Chapter 7).

#2. Looking uphill, this relatively steep tread is a former wagon road used for decades by several farms and ranches over a century ago. Given the site context, if the tread weren't so compacted by horse and wagon traffic, it would have eroded much more than this.

#4. Water channels from the rain that just ended [immediate foreground] will be obliterated by the first few visitors.

#5. [Foreground] Tread that would normally be sunken below the ground plane by compaction and displacement is instead level with the ground plane. This is because the sunken tread was filled by deposition from the eroding trail uphill [background].

CHAPTER 6

Tread Materials

To effectively design a trail, we must predict how *every* trail tread will behave. We've seen some of that in compaction, displacement, and erosion. Now it's time to examine how different tread materials—soil, rock, etc.—act under trail use in wet and dry conditions. This chapter discusses one core concept: tread texture.

Tread Texture

How can we describe tread components of soil and rock in simple terms? A way which works well for our purposes is examining how tread particles of different *sizes* behave. Soil scientists use the term "soil separates" to describe components of soil differentiated by size:

Clay	< 0.002 mm in size (extremely fine-grained particles)
Silt	0.002 to 0.05 mm
Sand	0.05 to 2.0 mm
Gravel	2 to 75 mm (from sand to 3")
Cobbles	75 to 250 mm (3" to 10")
Stones	250 to 600 mm (10" to 24")
Boulders	> 600 mm (> 24")

The Soil Triangle, above right, shows the relationships between the three smallest soil separate classes—clay, silt, and sand—and defines ranges for blends including loams.

Fortunately, most soils are blends of multiple separates. Why fortunately? Because each soil separate has definite strengths and weaknesses. **Blends of multiple separates tend to have the strengths of each of its components. The more separates a given tread has, the more likely it is that the strength of one compensates for the weakness of another. Hence the most desirable trail tread is a combination of as many separates as feasible for the trail type and purpose.** The term "texture" refers to the combination of soil separates.

Generally, though, we have to take what the site provides, working around weaknesses through design.

The table on the next two pages compares nine soil separates and textures including crushed stone and humus (organic soil) in terms of their properties and their behavior in trail treads. *Note that each separate exhibits its strengths and weaknesses in proportion to its amount in any given trail tread.* Other factors also exist, of course, but tread texture goes a long way toward explaining and predicting tread behavior.

***Guide for Textural Classification in Soil Families** ("Soil Triangle")*

Most soils are combinations of the soil separates clay, silt, and sand. While it isn't necessary to know the exact textural blend of trail tread soils, knowing the relationship between different blends is helpful for predicting tread behavior.

A few soil samples, many from one trail in a geologically diverse area.

45

Textures and Behaviors of Common Tread Materials

Name & Particle Size	Properties & Characteristics	Behavior in Trail Tread
Clay (0.002 mm) 	• Very fine textured. Pure clay doesn't feel gritty. Makes very fine dust. Sticky when wet. • Strong binder because of very strong ionic (electric) charges: a strongly held inner layer and a weaker outer layer. Inner layer and tremendous surface area attracts and holds water molecules on all sides. Outer layer charges "short out" when wet, allowing plate-like molecules to slide past each other (very slippery mud). • Clay particles have been chemically altered, i.e. not just broken rock.	• When compacted and dry, clays are extremely hard and resistant to displacement and erosion. • Will pond water for long periods. • Extremely slippery when wet, especially if saturated, uncompacted, or built up from fill. • Some clays expand when wet and leave shrinkage cracks when drying. • Clay is a good binder and is hence desirable in combination with other tread textures. • Pure clay tread is not desirable because it's too slippery when wet.
Silt (0.002 to 0.05 mm) 	• Fine- to medium-textured sediment from broken rock. Makes fine dust. • Larger than clay particles but smaller than sand, feels slightly flour-like. • Some silts are natural binders (raggedly shaped small particles with electrically charged edges) that can form very firm trail treads once compacted and when dry. • Soils with larger silt particles become less muddy than with fine-grained silts.	• Smooth and solid tread when dry, can be soupy and slippery when wet. • Silt compacts well but resists displacement and erosion less than clay. • Puddles will drain slowly. • Natural binding varies from very little to very strong. Particles with sharp edges have more binding than rounded particles. Particles with a variety of sizes bind better than particles all of one size. • Combining silt with other textures increases displacement and erosion resistance.
Sand (0.05 to 2.0 mm) 	• Coarse textured broken rock. Feels very gritty. Individual particles are clearly visible. • Very well drained, not muddy when wet. • No ionic charge, no natural binders, hence little resistance to displacement and erosion. • Due to increased mechanical interlock, sharp-grained sands resist displacement and erosion more than rounded grains.	• Sand doesn't sink much with compaction, but it doesn't harden much, either. • Pure sand treads are loose, easily displaced, and easily eroded. This makes it almost impossible to hold precise, subtle tread shapes such as outslope in pure sand. • Puddles drain relatively quickly and do not become muddy. • Pure sand tread is undesirable. But as part of a mix, sand adds drainage and compaction resistance.
Loam (<0.002 to 2.0 mm) 	• Loam is a mix of clay, silt, and sand (see Soil Triangle on previous page). • Most soils are loams. • Loam has properties of clay, silt, and sand. It has the binding of clay, the intermediate binding and structure of silt, and the compaction resistance and drainage of sand in the approximate proportions of their quantities in loam. • Beneficial qualities of each particle size tend to overcome the detrimental qualities of the others.	• Loam is smooth, firm, and stable on treads when dry. • Although any loam can be muddy, the higher the sand content, the more it resists being slippery and muddy when wet. • The more binding that clay and silt provide, the harder it becomes when compacted and the more it resists displacement and erosion. • A loam with a relatively even mix of clay, silt, and sand works better on treads than clay alone, silt alone, or sand alone.

Continued on next page

Name & Particle Size	Properties & Characteristics	Behavior in Trail Tread
Gravel (2 mm to 3") 	• Broken rock without binders. • Needs smaller particles, dust and compaction to fill voids and provide binding. • Larger particles increase bearing strength by providing a weight-bearing and weight-distributing "skeleton" in soil. • Angular particles have much more mechanical interlock and stability against shear forces than rounded particles. • Larger particles resist erosion through weight rather than binding.	• Increases bearing strength and load distribution. Doesn't get muddy. • Doesn't sink much with compaction. • The larger the particles, the more it can resist displacement and erosion. • Larger particles provide a wear surface that can protect smaller particles between them from displacement. • If all particles are the same size (graded or washed), gravel will have little resistance to displacement. • Becomes more stable when mixed with smaller particles, including clay.
Cobbles (3" to 10") **Stones (10" to 24")**	• Rock without binders. • Needs smaller particles, dust and compaction to fill voids and provide binding. • Provides even more weight distribution, weight-bearing strength, and displacement and erosion resistance than gravel. • Larger particles greatly increase bearing strength by providing a weight-bearing and weight-distributing "skeleton" in soil. • Rounded-edge rocks are friendlier to visitors than jagged rocks.	• Same benefits as gravel but with increased resistance to displacement and erosion. • Provides excellent wear surface for high displacement and heavy modalities, especially OHVs. • Superior resistance to shape change by compaction, displacement, and erosion. • Rough, bumpy tread surface. Visitors may try to avoid larger stones protruding from the tread. If smooth soil is adjacent, visitors may try to travel on it instead.
Crushed Stone (size varies) 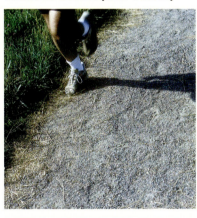	• Rock mechanically crushed from quarried stone or river deposits. • Many names including crusher fines, road base, aggregate. • For trail use, crushed stone should contain a full range of particle sizes from rock dust to a specified maximum screen size (i.e., size of screen that all particles must pass through). • Rock dust is essential to provide binding and stability. • Displacement resistance is largely from inertia of heavier particle weight. Hence the heavier the parent stone, the better for trails, i.e., granite is better than limestone. • *Book 2 has more crushed stone info.*	• Performance varies greatly depending on parent stone, particle size mix, and trail use. Finer mixes are smoother, coarser mixes resist erosion better through particle weight and mechanical interlock. • Surface may be loose. Can be dusty. • Resists shape change by compaction. • Moderate resistance to displacement. Resistance increases with larger maximum particle size, heavier parent stone weight, and good binder mix. • Does not get muddy. Can be stable under ponded water if previously well-compacted and left undisturbed. • Easily eroded by even small flows of fast-moving water.
Humus (organic soil, no size) 	• Organic product of vegetative decay. • Dark, spongy top soil layer in aerobic soils—essentially an advanced compost. • No binders, light in weight, very little mineral content, absorbs water well. • Very fluffy with many air spaces, compacts down to a thin layer over time as air spaces are crushed.	• Thick layers of humus compact easily, creating a deep tread rut even with little trail use. • Easily displaced and eroded unless bound by roots. Easily floats and easily washed away. • Too uncohesive to hold a sharp edge or subtle shape such as outslope. • Not recommended as tread material unless it is used only on lightly used trails with shallow grades, is bound by roots, and is protected from splash erosion (see page 55).

A Simple Field Test for Identifying Clay, Silt, Sand, and Loam

This simple field test helps identify the relative proportions of the finer textures in a soil:

[Flowchart: START — Place approximately 2 tsp. soil in palm. Add water dropwise and knead soil to break down all aggregates. Soil is at proper consistency when plastic and moldable like moist putty. → Does soil remain in a ball when squeezed? NO → Is soil too dry? YES → Add dry soil to soak up water; NO → Is soil too wet? YES (→ add dry soil); NO → SAND. YES → Place ball of soil between thumb and forefinger, gently pushing the soil with thumb, squeezing it upward into a ribbon. Form a ribbon of uniform thickness and width. Allow the ribbon to emerge and extend over forefinger, breaking from its own weight. Does soil form a ribbon? NO → LOAMY SAND; YES → Excessively wet a small pinch of soil in palm and rub with forefinger into a ribbon → Does soil make a weak ribbon < 1" long before it breaks? YES → (gritty? YES SANDY LOAM; neither gritty nor smooth predominantly? YES LOAM; smooth? YES SILTY LOAM). NO → Does soil make a medium ribbon 1–2" long before it breaks? YES → (gritty? YES SANDY CLAY LOAM; neither? YES CLAY LOAM; smooth? YES SILTY CLAY LOAM). NO → Does soil make a strong ribbon > 2" long before it breaks? YES → (gritty? YES SANDY CLAY; neither? YES CLAY; smooth? YES SILTY CLAY).]

Tread Texture and Tread Performance

The table on the two preceding pages makes several ideas clear:

1. **Knowing tread texture helps you predict how a tread accommodates physical forces.** Each texture class has definite properties that act differently under compaction, displacement, and erosion forces in wet and dry conditions. Knowing how tread textures behave and interact largely tell you how a given tread will handle forces acting on it. Texture properties also indicate ease of use, or lack of it, that may prompt visitors to use or avoid the tread.

2. **Binders bind fine textures to each other.** Clays and some silts hold ionic (electrical) charges that make them stick together. Water, however, shorts out some of the attraction, causing plasticity and mud when fine textures are moist and saturated, respectively.

3. **The more soil separates a tread has, the stronger it is.** If you've ever been on them, you probably noticed that treads of pure sand are too loose, that pure silt is dusty and may be mucky when wet, that pure clay is hard when dry but terribly slippery when wet, and that gravel or stones without fines remain loose and shifty. On the other hand, you've probably seen stable, durable treads with a full range of separates from clay to stones that become hard with compaction, remain firm when wet, and resist displacement and erosion well. As mentioned previously, the difference is the mix and the fact that the strengths of each separate help compensate for the weaknesses of others.

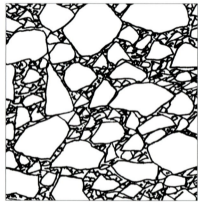

A wide range of particle sizes from clay and silt and larger makes the best soil for trail tread. The smallest particles are binders, larger particles better resist displacement and erosion while providing strength in wet conditions, and medium particles of all sizes add structure that helps stabilize the tread. Compaction greatly strengthens such treads by eliminating spaces and improving binding.

4. The larger the largest tread particles are, the more the tread can resist tread shape change through compaction, displacement and erosion. Through their weight, size, and structural integrity, large particles provide a displacement- and erosion-resistant wearing surface. To be stable, though, fine particles must surround and fill all voids between large particles.

With its full range from fines to cobbles, this well-compacted tread is highly resistant to displacement and erosion. In fact, this is part of a road that supports frequent truck traffic. The round disk above the center is an American quarter coin for scale.

Fine-textured treads can displace easily when wet, especially under modalities that concentrate weight in a small area.

Tread Hardening

If a tread cannot sufficiently resist compaction, displacement, and/or erosion, hardening the tread may be appropriate. As the above photo shows, augmenting or replacing native soil with crushed stone, gravel, cobbles, and even stones can firm or harden most treads.

Book 2, *Shaping Natural Surface Trails by Design: Key Patterns for Forming Sustainable, Enjoyable Trails,* discusses hardening with stony material, other tread hardening methods, and alternatives to hardening.

Testing Performance of Tread Materials

The easy-to-perform test on the next page simulates the effects of compaction, displacement, erosion, and water on any tread material. You only need a small sample of tread material and you can do the test in the field or at your office. In the field, I collect tread material samples with sturdy "zipper"-style plastic freezer bags and a small gardener's shovel.

The test is highly recommended for soils and tread materials that are unfamiliar to you and for imported materials (soil, crushed stone, gravel, etc.). Save yourself tons of grief by testing materials for suitability before you commit to them.*

* Especially with imported crushed stone and gravel, I've seen numerous treads that had to be redone or extensively repaired because the material didn't perform as expected. The test on the next page could have predicted all of those failures.

How To Estimate the Performance of Any Tread Material

A small sample of tread material will act just like the tread itself. If you're in doubt as to how a new trail tread or particular tread material will perform, conduct this simple test. The test simulates tread compaction, displacement, and erosion in wet and dry conditions with no measuring, no math, no lab tests, and no cost. The photos below show a clay loam.

Warning: This easy test eliminates excuses for **not** knowing in advance how a tread will perform.

You will need: material sample, mold, plastic wrap, hammer-like tool, screwdriver-like tool, pourable water.

1. If the sample isn't already moist, add water [shown above] until particles stick together when compacted. Heap the moist sample in a sturdy, can-like, smooth walled mold. A white 4-inch PVC drain cap is shown here. Line the mold with kitchen plastic wrap to make steps 7-9 easier.

2. Compact the sample with a hammer as hard as you can. If sample is too wet, let it dry a bit first. Form a shallow depression in the center. Let it dry for several days—in a warm and/or sunny place if possible—indoors or out. The drier, the better.

3. After the sample is thoroughly dry, use a sharp-ended tool to scrape and chip at the surface, then gouge into the surface. Be brutal. The firmness or looseness you see here will also happen on the trail. This tests the resistance to displacement of the compacted tread.

4. Fill the center depression with about 1/8" of water and note the absorption rate. This indicates tread permeability and runoff rates. The slower the absorption, the more runoff the tread itself will create. Let water sink in until none is left on the surface.

5. Scrape and gouge the wet portion of the sample. Note how similarly or differently the sample acts compared to step 3. This simulates trail use on a moist tread. Again, what happens here will happen on the trail.

6. Tilt the container a bit, then use a pressurized hose or pour a stream of water from some height to simulate an erosive water flow. [In the photo, water is coming down from the top center]. Erosion that happens here is what to expect on the trail.

Steps 7-9 are an optional extension
7. By holding onto the plastic liner and pulling, try to remove the sample from the mold in one piece (requires a smooth-sided mold). If the sample falls apart, it will displace easily on the trail.

8. If sample is solid, break off a chunk to see how deep the compaction went and how firm it is. The deeper and firmer it is compacted and hardened, the better. This sample was very hard and had to be hit with tools to break it.

9. As in step 6, simulate an erosive water flow on a level or near-level surface. Note how readily chunks break down and particles wash away. Note the difference in erosion between previously wet and previously dry portions.

● CHAPTER 7
Tread Watersheds

In Chapter 5, we discussed how compaction, displacement, and erosion tend to reshape trail tread. Chapter 6 explained how physical forces and water tend to reshape various tread textures.

In this chapter, we look at water and the tread within its site context—where water comes from, how it moves across and along trail tread, and how water movement tends to affect the tread. To discuss this context, I've coined the core concept *tread watershed*.

Tread Watershed

A watershed is the land area that drains into a given water body or channel. A *tread watershed*, however, is a bit different. A tread watershed is the trail tread between a local high point (crest) and the next local low point (dip), plus the land area that drains onto this tread segment:

Tread watershed boundaries

Each tread watershed is assumed to drain through the dip at its lowest end

Tread watershed height is from the downhill edge of the tread up to the topographic top for drainage

Tread watersheds catch water from the site above the tread plus rain, snow, and seepage landing on the tread itself

Length of a tread watershed is the tread length between a local high point (crest) and the next local low point (dip) in the tread. **Crest and dip locations may or may not be tied to site topography.**

The tread watershed concept looks at a trail as a series of waterways. Why? Because even if a tread is outsloped, compaction, displacement, and erosion tend to defeat outslope (Chapter 5). If and/or when outslope fails—or if outslope wasn't originally present—the tread *becomes* a waterway when water follows the tread rut. At that point, the tread drains solely through its dips.

Like a conventional watershed, the height of a tread watershed is from the lower edge of the tread upward to the top of its topographic basin. Tread watershed length, however, is set by the locations of local crests and dips in the tread.

What about outslope?
*An outsloped tread forms a theoretically infinite number of tread watersheds of near-zero length. This would be perfect if we could count on tread maintaining its outsloped shape. Yet, as we've seen, all human and natural forces tend to work **against** outslope, making outslope difficult to maintain. Managing tread watersheds is the next best thing to sustainable outslope.*

Mount St. Helens

BOTH PHOTOS Mount St. Helens National Volcanic Monument, Washington, 19 years after the devastating 1980 blast and eruption. Formerly forested, this ridge is now denuded and covered with a highly erosive layer of ash. Blast-felled trees point to Mount St. Helens on the left. In the top photo, note how eroding gullies formed just below the top of the ridge.

The new, post-blast trail in both photos carefully traverses many tread watersheds. Orange dashed lines [above] show watershed boundaries above crests while gullies mark the dip boundaries. Each of the many actively eroding drainages is crossed as a dip in the tread [left]. This allows each drainage channel to cross the tread instead of being intercepted and channeled down the trail. Limiting the width of tread watersheds—by intentionally forming dips and crests into the tread alignment—is the only way to form a sustainable natural surface tread in highly erosive conditions such as this.

Crest and dip locations, however, are our decision. They may or may not be tied to topography. For instance, in the drawing on page 51, the dip on the left is tied to a natural drainage while the dip on the right has no explicit relation to site topography. The closer together the tread crests and dips are, the smaller the tread watersheds. Hence **tread watershed sizes and locations are partly a human choice determined by the exact tread location in the local site.** This is true no matter how the tread came to exist.

Adding drainage dips, waterbars, and other drainage devices to trails can reduce the length of tread watersheds (see photo at right). These devices do the job but require extensive construction and ongoing maintenance (reshaping) to keep them functioning. For existing trails, they are often the only option for limiting erosion. For new trails, though, it's better to plan appropriately small tread watersheds into the alignment from the beginning. This, along with drainage devices, is discussed in Book 2.

Note that the tread itself is an important part of its tread watershed. Rain and snow falling directly on the tread affects the tread along with water draining down from the area above. In fact, as you'll see in the following discussion, the tread itself is a major "source" of runoff water.

Looking downhill, this series of waterbars was added only months after initial trail construction. They were added to form smaller, more sustainable tread watersheds after erosion damaged the original long, steep tread. As discussed in this chapter, the damage was predictable.

Note: I was asked to design this trail but declined since the site cannot sustainably support as steep a trail as the land manager wanted. The manager shaped an excessively steep trail anyway, the photo shows part of the result, and future generations are now condemned to extra work to keep it sustainable.

Tread Watershed Factors

Tread watersheds subdivide a trail into segments where we can analyze water flow separately. That's very useful in itself. But our greatest concern is the *performance* of the tread in each tread watershed. We care about water sources, how much water there is, how fast it moves, where it flows, and how well the tread can accommodate it. We also need to evaluate performance of existing treads and predict performance of new treads before they even exist.

We can use twelve factors to analyze and predict tread watershed performance:

- Tread watershed size
- Watershed slope
- Runoff potential
- Splash erosion
- Tread width
- Weather, climate, and microclimate
- Water sources
- Tread texture
- Trail use (compaction and displacement)
- Tread grade
- Tread length
- Dip sustainability

Each factor is simple in itself. Some are core concepts from preceding chapters (physical forces and tread texture). Yet because all twelve factors affect tread sustainability, they are some of the most important keys of natural surface trail design.

Each factor is discussed below.

Tread Watershed Size

The larger the tread watershed, the more water it collects from rain and snow. Small tread watersheds help limit how much water can potentially reach the tread.

A large tread watershed, mostly from length but partly from height. The tread, with a 14% grade, is currently outsloped but has a steep watershed slope and high runoff potential. In the future, it will likely need waterbars or drainage dips to form shorter tread watersheds.

Watershed Slope

The steeper the slope above the tread, the more likely that water will run off of it and the faster that the water can move. Hence a steep slope will likely produce more runoff than a shallower slope with the same runoff potential.

Level and near-level tread watersheds pose a different problem. Because compaction and displacement almost always act to lower the surface of natural surface treads (see page 38), an "unimproved" tread tends to become the lowest place around. The tread can then become a ditch that channels water, ponds water, and is difficult to drain. This is the main reason why level and near-level sites are problematic for natural surface treads. Book 2 discusses the problem and solutions in detail.

Where slopes are minimal, it's difficult to drain treads to the side. Fresh erosion from a quick melting late spring snow shows as the dark strip draining down the center of this sandy loam tread.

Parallel and braided treads occur when (1) there is minimal sideslope and (2) visitors seek greater travel efficiency by trying to avoid wet, muddy, low-lying tread ruts by traveling on higher, less muddy areas.

Runoff Potential

Tree canopy, vegetative cover, and forest litter in the tread watershed help determine how much runoff is generated by rainfall:

Tread watershed attribute	Runoff potential
Dense tree canopy	Low
Thick forest litter on ground	Low
Moderate tree canopy/dense shrub canopy	Medium
Sparse tree/brush canopy	High
Meadow/grassland	High
Exposed bare soil	High-very high
Clay, impervious, or rocky soil	Very high
Rock	Extreme
Hardened surface (pavement, parking)	Extreme

Rock produces 100% runoff. Tread below rock outcrops and rocky slopes can be hit with torrents of runoff in heavy rains. This tread requires risers and frequent waterbars to reduce tread grade and tread watershed sizes, respectively.

Features that absorb water or sharply reduce the speed of raindrop impact with the earth, such as thick forest litter, help reduce runoff. Bare soil, rock, and other hard surfaces produce the most runoff. Mixed attributes, such as thick forest litter and rocks, tend to produce runoff in the proportions of their presence.

Runoff speed generally increases with runoff potential. Low-runoff situations generally have slow runoff such as the slow seeping from a dense forest with thick forest litter. High and extreme runoff from surfaces such as from rock typically happens quickly, concentrating moderate rains into fast streams and heavy rains into torrents.

Note, however, that torrential rains can produce high runoff regardless of watershed attributes.

Splash Erosion

Have you ever been out in a stinging hard rain that tempted you to take shelter under a tree? Trees protect trail tread, too. Where tree canopy extends over the tread, foliage and branches take the direct hit of raindrops. Water then drips gently to the tread below.

Without a canopy, raindrops strike the tread at full speed. Their impact displaces and suspends dust and small particles

Close-up of gravelly tread after erosion. Splash erosion dislodged fine particles that washed away in runoff erosion.

that can then easily wash away, especially in a heavy rain. This is called splash erosion. Over time, splash erosion increases local tread erosion compared to similar tread protected by a tree canopy.

Hence treads exposed to the sky will experience splash erosion in addition to any other erosion. And the steeper the tread grade, the more that splash erosion contributes to runoff erosion.

Splash erosion contributes to runoff erosion on this exposed tread. Note how gravel, foreground, protected the soil beneath it from recent hard rain.

Tread watershed planning should take splash erosion into account and shorten tread watersheds as needed.

Tread Width

Tread width is critically important. As discussed on page 36, tread compaction causes most of the rain that falls directly on the tread to run off even in a light rain. Hence the wider the tread, the more surface area it has and the more runoff it generates. In fact, in many runoff events, all or most of the water on the tread is from rain that fell directly on it. In these cases, a narrow tread helps to limit runoff more than having a small tread watershed.

In considering and planning tread watersheds, plan for 95-100% runoff from the tread itself.

Weather, Climate, and Microclimate

Even if uncommon, in regions where storms or hurricanes can dump 1", 2", or more rain per hour (or in minutes), you must assume that intense rains *will* occur someday. Such rains will overwhelm the runoff-buffering capability of thick tree canopies and thick forest litter, potentially delivering a cascade onto the tread. Limiting tread watershed size is the only way to limit runoff from torrential rains.

Rapid snowmelt, especially on south-facing slopes, can also spawn erosive flows. I saw a trail wash out when 8 inches of wet spring snow melted in a few hours on a rocky, south-facing slope.

In higher elevations and northern latitudes, north-facing slopes are often much cooler and wetter than south-facing slopes. This affects the speed of snowmelt, temperature, humidity, vegetation, and the length of time a tread is likely to remain wet. The bottoms of deep ravines may also be cooler and wetter.

Trail planning should take the possibility of severe weather, overall climate, seasons of use, and microclimates into account.

Tread width: The wider the tread, the more water it catches and carries from rain and snow. Plus, additional width of bare soil exposes more soil to potential erosion. For maximum sustainability, natural surface treads should be as narrow as feasible.

Water Sources

All sources of water in a tread watershed need to be accommodated. Water sources may include:

- Rainfall and/or snowfall. These affect all tread watersheds
- Ephemeral drainages, typically short-term rain or snowmelt runoff
- Intermittent or seasonal drainages
- Perennial drainages (streams)
- Springs
- Seeps (seasonal or perennial)
- Hanging or perched water table. A tread traversing a slope with an impervious layer a few feet below the surface may have water below the surface but above the impervious layer. Cutting into the slope may cut into the wet level and release subsurface water into the trail cut
- Local flooding from a channel or floodplain adjacent to the tread

A seasonal (springtime) seep at the base of a slope adds water directly to this tread. Seeps are common at the bases of large rock outcrops and rocky slopes.

We've established that natural surface treads tend to catch and channel rain and non-point-source runoff because often we can't prevent it (Chapter 5). But most of the above water sources flow in defined channels or areas, and most repeat in the same locations. Using techniques discussed in Book 2, we need to accommodate water sources by letting them flow across, under, or beside the tread—but not directly along the tread itself except by intention.

When planning or evaluating tread, watch for any sign of water in the form of defined channels, unusually wet soils, unusually lush vegetation, or presence of water-seeking plant or tree species. The presence or absence of water should be a major planning factor for trail location, tread watershed location, and tread formation.

Tread Texture

As discussed in Chapter 6, some tread textures are better than others at resisting erosion and maintaining their shape when wet:

- The most erosion-resistant treads have a well-compacted mix of all textures including gravel and sometimes larger particles.
- Graded (particles all of one size) materials larger than clay lack binding and tend to erode, including gravels.
- Compacted clay and some compacted silts can be quite erosion-resistant but are often very slippery when wet and structureless when uncompacted.
- Clays, especially expansive clays, can hold water longer than coarser soil textures, causing clay-rich treads to remain wet longer.
- Angular particles that tend to mechanically interlock are more erosion-resistant than round particles.
- Particles loosened by displacement are the first to wash away.

For sustainability, water flow and water speed must be limited on tread textures with low erosion resistance.

Water flowing down the tread—from a new spring that emerged in the middle of the tread in a limestone canyon—is similar to runoff during a heavy rain. The well-compacted, rocky tread suffered little damage from this flow, but many other treads would be heavily damaged in a similar situation. How any tread performs in wet conditions and under runoff depends largely on tread texture.

Trail Use (Compaction and Displacement)

As discussed in Chapter 5, the type and amount of trail use affects the speed and degree of tread shape change through compaction and displacement.

On page 41, we discussed how specific modalities tend to affect the tread through compaction and displacement. High-displacement modalities are more likely to reshape the tread by causing loss of outslope, superelevating curves (wheeled modalities), knocking the tops off of crests through displacement, and filling in dips through displacement (also see "Dip Sustainability" on page 60).

Higher amounts of trail use, especially by high-displacement modalities, can reshape the tread faster and/or more.

In planning and evaluating tread watersheds, consider the effects of compaction and displacement exerted by the type and amount of trail use. *Be especially careful if you expect outslope to be maintainable or if you expect high amounts of high-displacement use.*

Risk of tread being reshaped solely through trail use

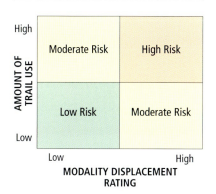

Tread Grade and Tread Length

The steeper the tread grade, the more likely it is to erode. Erosion tendency, however, increases exponentially with grade, i.e. small increases in grade greatly increase erosion potential.

Tread length is also critical. A tread segment running at a 20% grade for 2 feet has nowhere near the erosion risk of tread running at 20% for 100 feet.

What everyone wants to know is, "How long and how steep can a tread segment be before it becomes unsustainable?" That depends on the grade, length, tread texture, type and amount of trail use, amount of water likely to be on the tread, and all the other tread watershed factors. **It's impossible to provide absolute numbers,** but we can make approximate estimates for hypothetical examples based on empirical observations of grade, length, and tread texture.

For instance, the table and graph on the next page suggest maximum tread lengths at different grades for several tread textures. I determined lengths based on informal observations of many trails over a period of years, combined with predictions. The table notes the many specific assumptions underlying the length values, and again, **these are hypothetical values.** Also note that the stated maximum tread lengths were chosen to require minimal maintenance in contexts with moderately high erosion risk. Tread watersheds with less erosion risk enable longer tread lengths. And, in any context, lengths can be "pushed"—but if so, they tend to become more risky for erosion and/or require more maintenance to maintain tread shape.

The overall pattern, though, is far more important than individual numbers. Note how quickly the lengths shrink as the grade increases.

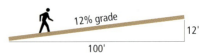

Grade measured as a percentage is the number of feet climbed in 100'. Here, climbing 12' in 100 horizontal feet is a 12% grade. You can use a clinometer or Abney level to measure grades.

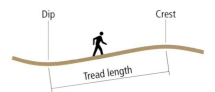

Too steep for too long—on sideslope
High displacement horse use in sandy soil, plus too much water on a tread that is too long and too steep, resulted in this trench traversing a slope. Tread watersheds need to be smaller.

Too steep for too long—on fall line
The tread in the foreground eroded to 12' wide and 3' deep from going straight down the slope (fall line) in clay soil. Fall line alignments are virtually impossible to drain (Book 2 explains why).

Too steep for too long—nearly level
This crushed stone tread washed out from too much water moving too fast for it. Even when tread is nearly level, limiting erosion is a matter of forming sufficiently small tread watersheds.

The graphed values below clearly illustrate exponential change with small changes in grade. **All tread textures have similar patterns.**

You're strongly encouraged to customize tread grades and lengths for your own trails based on your own observations. Make sure that you take all of the tread watershed factors into account or your values won't correspond to the real world.

Approximate Hypothetical Maximum Tread Lengths by Tread Texture, with Specific Assumptions

Tread length is the distance between a tread crest and the adjacent dip that drains the segment between them.

Assumptions:
- Most tread watershed water drains down the tread and through the dip at the lower end (i.e., sunken tread with little side drainage).
- **Although erosion will still occur even with these values—especially in extreme runoff events**—these tread length values are designed to require minimal tread maintenance and to minimize tread shape change through erosion.
- Tread is **well-compacted** and about 30" wide.
- Trail has moderate use with moderate displacement (hiking).
- Tread watershed has moderate runoff potential.
- No tree canopy (high splash erosion).
- Downpours are likely only 1-3 times each year (climates with more extreme rain should use shorter tread watersheds).
- No water sources exist besides than rain and runoff.
- Note that these values are intended only as a conceptual guide. YOUR CONDITIONS AND RESULTS WILL VARY.

Tread texture	0%*	2%	4%	6%	8%	10%	12%	14%	16%	18%	20%
Clay loam with high quantity of gravels, cobbles, and stones	215'	160'	120'	90'	67'	50'	35'	24'	16'	10'	5'
Gravelly clay	180'	132'	96'	69'	49'	34'	22'	14'	8'	4'	
Loam with high quantity of gravel and cobbles	160'	117'	83'	57'	39'	26'	17'	10'	6'	3'	
Clay†	145'	104'	74'	51'	34'	22'	13'	7'	4'		
Loam	135'	90'	57'	37'	23'	14'	8'	4'			
Crushed granite or crushed limestone, angular particles, 0.75"-minus, 5" thick	125'	78'	49'	30'	17'	9'	5'				
Organic soil	110'	68'	39'	22'	12'	6'					
Sand	100'	55'	30'	16'	8'	3'					

* Unless it is sustainably pitched to drain to the side, no tread should have a 0% grade. The 0% figures are listed as an upper dip spacing limit for grades above 0% and below 2%.

† Although compacted pure clay can be cohesive even on steep grades, it is generally too slippery when wet to be practical.

When the above values are plotted, their overall patterns are clear:

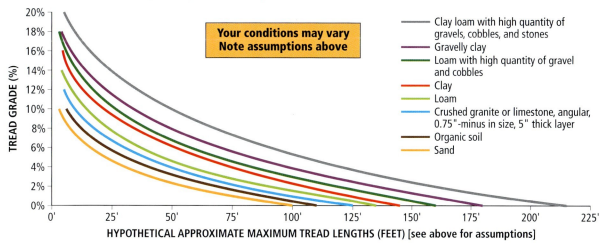

Dip Sustainability

While "tread dip" is a simple concept, constructed dips (waterbars, drainage dips, etc.) and dips formed by tread alignment can have many problems and fail in many ways. Problems include scouring, slow or poor drainage, clogging, ponding water, and loss of the entire dip—all of which can occur in various ways and combinations. Hence, dips are surprisingly complex.

The most sustainable dips, however, have six simple, interrelated characteristics:

1. **Minimal sediment flow:** If a dip carries sediment, that's usually the tread eroding. Ideally, tread watersheds are small enough and tread watershed factors are such that minimal amounts of displaced and eroded material flow through dips. We expect more material to flow through dips in extreme runoff events. But if a dip scours deeper or carries too much material in anything less than extreme events, the tread watershed is too large.

2. **Quick drainage to the side:** Water should flow through and out of the dip faster than it flows into it. This is easiest when water drains down a slope away from the tread. *The more the outflow slope exceeds the tread grade, the better.*

    What happens if the outflow slope is **less** than the inflow slope? Water slows down in the outflow and drops part or all of its sediment load. This may clog the dip in time and exacerbates #4 and #5 below.

3. **Wide outflow:** Wide outflow channels help resist clogging.

4. **Tread compaction and displacement resistance:** Tread compaction and displacement can lower the tread through the dip, forming a berm (dam) across the outflow and/or ponding water in the dip. In addition, on steeper treads, displacement and gravity can roll tread material down the tread until it lodges in a dip, fills the dip, or clogs the outflow at the lower edge of the tread.

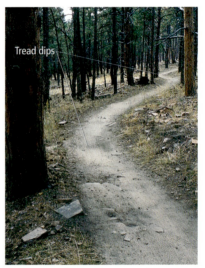

Ideal tread dips. At best, tread dips are formed by tread alignment rather than constructed, drain quickly to the side, have a wide outflow, carry sediment only during major runoff events, and are large enough to retain their shape despite tread reshaping by trail use.

Problem with #1: Although this constructed drainage dip drains the tread well, too much tread material is flowing through it. Look how much tread material has been lost above the dip. The tread watershed should be smaller (i.e. another dip is needed) or the tread grade should be less steep.

Problems with #2 and #4: This constructed drainage dip (the puddle) is supposed to drain to the right, yet there isn't much slope in that direction (#2). ATV use displaced the sandy clay tread in the dip, creating a berm across the outflow (#4), blocking drainage. Trail use also splashes (displaces/erodes) tread material out with the water, deepening the pond into a mudhole (#4).

Problems with #2, #3, #4, #5, and #6: *On this steep tread, the inflow to this flexible waterbar is steeper than the outflow, allowing sediment to deposit in the dip (#2). The narrow outflow contributes to rapid clogging (#3). Displaced silt from the steep, silt tread also tends to roll down the tread, piling up in the dip where displacement pushes it to the lower edge (#4). And both the dip and crest are too small for this tread watershed context with its steep, silty tread (#5, #6).*

Problems with #2 and #6: *This large waterbar drains the tread—but the price is high. The crest, while substantial, is almost a barrier across the trail, requiring a high step (#6). It tempts visitors to try to bypass it, necessitating a stone extension on the uphill side. The waterbar should also be angled more so it doesn't slow water speed as much as it does now (#2). Both sustainability and the trail experience would benefit from more frequently spaced, smaller waterbars.*

Note that long-term ponding can turn into a mudhole, especially if quick drainage to the side (#2) is lacking.

5. **Sufficiently sized dip:** The dip needs to be deep and wide enough to function even if displaced or eroded material deposits in the dip or outflow. If excessive amounts of material deposit in a dip, the tread watershed is too large and/or the tread should be hardened.

6. **Substantial crest:** The crest that backs the dip needs to fully accommodate trail use. It must be sufficiently large and/or strong to avoid being knocked away (displaced). Each crest must also be sufficiently easy to use that visitors don't bypass it and form a new water bypass channel in the process. Substantial crests are critical for waterbars and drainage dips because if their crests are lost, often the dips are lost, too.

Dips with all six characteristics require little to no maintenance. In contrast, the more of the above characteristics that a given dip lacks, the more maintenance it requires and the more likely it is to develop problems, fail under extreme conditions, or fail between maintenance intervals.

At worst, the failure of a dip or crest merges adjacent tread watersheds and combines their water, potentially overwhelming the lower tread watershed(s) with much more water than before. This can cause major tread damage with heavy runoff.

Bottom line: In planning and evaluating tread watersheds, ensure that the dip at the bottom can actually handle both water and the physical forces caused by trail use.

Book 2 covers dip sustainability in more detail, including shaping and maintenance details.

Problems with all 6 characteristics: *Three consecutive stone waterbars failed from too much sediment flow (#1), slow drainage to the side (#2), narrow outlets (#3), high tread displacement (#4), shallow dips (#5), and insubstantial crests (#6). Now that they've failed, three tread watersheds merged into one, accelerating trail erosion.*

Putting All the Factors Together

The table below summarizes the twelve tread watershed factors. Once you've learned them, you can assess them very quickly (Chapter 8). You'll find that tread watersheds and their factors concisely explain much of the how and why of natural surface trail treads.

To help learn tread watershed factors, visit your favorite trail and practice applying all twelve factors to each tread watershed. Look for problems and explain them. Also look for successful tread watersheds and explain them. Try to develop your own table of tread grades and lengths along with rules of thumb for common situations.

Quick Reference: Tread Erosion and Water Damage Risk

For a given tread watershed, the more factors that fall into the rightmost column below, the more carefully water must be accommodated to reduce risk of erosion and water damage:

Tread watershed factor	Decreased erosion & water damage risk	Increased erosion & water damage risk
Tread watershed size	Smaller tread watershed	Larger tread watershed
Watershed slope	Shallow slopes	Steeper slopes
Runoff potential	Low runoff potential (thick forest litter)	High runoff potential (little cover, rocky)
Splash erosion	Tree canopy over tread	Tread open to sky
Tread width	Narrower tread	Wider tread
Weather, climate, microclimate	Light rains only, slow snowmelt	Downpours, heavy snows, rapid snowmelt
Water sources	No water sources, constant and/or limited water sources, low water table, water easily anticipated and accommodated	Unpredictable or highly variable water sources, high water table, water not easily drained or accommodated, floodplain
Tread texture	Compacted tread surface that is not easily displaced, some larger particles/rocky content, dry or moderately dry tread	Easily displaced and/or graded materials (all one size), no larger particles, too many round particles, wet or saturated tread
Trail use (compaction & displacement)	Low trail use, low displacement modalities, low likelihood of tread crest/dip failure	High trail use, high displacement modalities, higher likelihood of tread crest/dip failure
Tread grade	Shallow grades	Steeper grades
Tread length	Shorter tread length	Longer tread length
Dip sustainability	Minimal sediment, quick drainage, wide outflow, minimal tread displacement, sufficient size	Too much sediment, slow drainage, narrow outflow, high displacement, insufficient size

Predicting the Future

With practice, you'll be able to predict the future of any tread watershed. In fact, this is how I became aware of these factors and their effects. I studied existing trails, made predictions of what I thought would happen when, then returned months and years later to see what actually happened. If something unanticipated occurred, I looked for the cause and factored it into future predictions. Using the factors presented here, I eventually reached the point where I could accurately predict both future trends and timelines for most tread watersheds.

To become proficient at natural surface trail design, start evaluating trails and predicting their future. That's the next chapter.

● CHAPTER 8

Trail Evaluation: Reading Trails Like a Book

Old Problems with Trail Evaluation

Exactly what do we evaluate? What do we measure? How do we describe what we see? Because we lacked standards for what and how to evaluate—and also lacked language to talk about the process and results—these questions often stymie trail evaluations.

To create standards and language, some try to use scientific-style monitoring with numeric measurements, baselines, and databases with rigid parameters and very limited applications. But while monitoring is useful for answering specific quantitative questions, it can't quickly evaluate the complex functioning of natural surface trails. And monitoring can't tell you **why** something is happening.

Easier Evaluation with the Foundation Level

Both the Foundation Level and trail evaluation are about the same thing: *the essentials of what works and what doesn't—and why—where the trail touches the ground.* Hence the eleven core concepts of the Foundation Level make excellent evaluation criteria:

- They can be used to evaluate trail and tread quickly, thoroughly, and accurately.
- They don't require numeric measurements, baselines, or databases (although you can use them if you like).
- They focus on cause and effect rather than states and conditions.
- They provide a concise set of what to examine.
- They provide concise language to capture what we see.
- They capture both visitor enjoyment and physical sustainability.
- They spotlight aspects of high and low performance.
- They can be used on both existing and proposed trails.
- With some training and practice, they facilitate a high degree of consistency from one evaluator to another.

The next page shows a one-page trail evaluation form for any natural surface trail. Although it's optimized for evaluating individual tread watersheds, you can use it on larger segments or even entire trails. Space is provided for predicting the future of the segment being evaluated as well as for specific comments and recommendations, so you can easily use it as a trail condition/maintenance request log as well. You can also add your own criteria for your particular situations or concerns.

Note that Foundation Level evaluation concentrates on what is readily apparent on the site or can be directly inferred from what is apparent (rainfall amounts, runoff, etc). Hence it does not include

Core Concepts

Natural shapes
Anchors
  • Edges
  • Gateways
Safety
Efficiency
Playfulness
Harmony
Compaction
Displacement
Erosion
Tread texture
Tread watersheds
  • Tread watershed size
  • Watershed slope
  • Runoff potential
  • Splash erosion
  • Tread width
  • Weather, climate, and microclimate
  • Water sources
  • Tread texture
  • Trail use (compaction and displacement)
  • Tread grade
  • Tread length
  • Dip sustainability

Quick Trail Evaluation Form—Foundation Level

Trail and tread are best evaluated in short sections. Here, "trail" generally means one tread watershed or a segment of a longer trail. In numeric scoring, 0 = not much or not well, 5 = very much or very well. Since trails tend to be "weakest link" situations, the overall score is the average of the lowest score in each category.

TRAIL/SEGMENT LOCATION: _____

Human Perception
Relative to its potential, how much does the trail incorporate/reflect/explore/integrate with the site through use of:

Natural shapes   0 1 2 3 4 5   Comment:_____
Anchors   0 1 2 3 4 5   Comment:_____

Category score *(lowest score in this category):* _____

Human Feelings
How much does the trail create/shape/promote contextually appropriate feelings of:

Safety   0 1 2 3 4 5   Comment:_____
Efficiency   0 1 2 3 4 5   Comment:_____
Playfulness   0 1 2 3 4 5   Comment:_____
Harmony   0 1 2 3 4 5   Comment:_____

Category score *(lowest score in this category):* _____

Physical Forces
How well does the tread retain its functional shape in the face of:

Compaction   0 1 2 3 4 5   Comment:_____
Displacement   0 1 2 3 4 5   Comment:_____
Erosion   0 1 2 3 4 5   Comment:_____

Category score *(lowest score in this category):* _____

Tread Material
How well does the tread texture help the tread retain its functional shape in the face of trail use, water, etc., when **dry?**
0 1 2 3 4 5   Comment:_____

How well does the tread texture help the tread retain its functional shape in the face of trail use, water, etc., when **wet?**
0 1 2 3 4 5   Comment:_____

Category score *(lowest score in this category):* _____

Tread Watershed
How much do tread watershed size; watershed slope; runoff potential; weather, climate, and microclimate; and water sources **all** seem limited to what the tread can sustainably accommodate?
0 1 2 3 4 5   Comment:_____

How well is the tread retaining its functional shape given the combination of splash erosion, tread width, tread texture, trail use, tread grade, and tread length?
0 1 2 3 4 5   Comment:_____

Is dip sustainability adequate both now and until the next scheduled maintenance? *(If in doubt, be conservative)*
0 1 2 3 4 5   Comment:_____

Category score *(lowest score in this category):* _____

▶ **OVERALL TRAIL/SEGMENT SCORE (average of the five category scores):** _____ out of 5.0

Prediction/Recommendation/Notes: _____
_____
_____

ecological considerations, management concepts, social contexts such as visitor conflicts, etc. Book 3, *Managing Trails by Design: Integrating Stewardship, Sustainability and the Trail Experience,* provides a more thorough trail evaluation process that examines the full trail context.

Evaluation form details

You are encouraged to use or modify this evaluation as you see fit.

Note that scoring is done per category. But because trails are a "weakest link" situation at the Foundation Level, the lowest individual score in each category determines the score for the entire category. This provides a more accurate gauge of what's actually happening.

The Tread Materials category has one item for when the tread is dry and one for when it's wet. Since treads composed primarily of clay, silt, and loams tend to rate poorly when wet, use this with discretion to help obtain a useful score for common trail use conditions. Avoiding trail use during spring melt or saturated conditions can greatly help reduce wet tread damage.

The "Prediction/Recommendation/Notes" field at the bottom is for the evaluator's prediction of what will likely occur given likely future trail use, maintenance, etc., as well as recommendations or notes of interest. This can also be used as a maintenance request log or to comment on the status or effectiveness of past maintenance efforts.

Evaluation as a Trail Design Exercise

Design attempts to solve problems. Evaluation examines how well design is working. Hence practicing evaluation will make you a better trail designer.

Although the Quick Trail Evaluation Form provides a good way to evaluate a trail quickly, expressing a trail in your own words from scratch is far better for learning. What works and what doesn't is clear if (1) you know where to look and (2) you can put the process and your findings into words. *The core concepts help you do this.*

Beginning on the next page, a series of evaluations of real-world trails from many contexts completes this chapter. These evaluations are learning tools to demonstrate

- how to apply the core concepts to many kinds of trails, and
- how thoroughly yet succinctly we can use the core concepts to evaluate trails.

Rather than simply demonstrate use of the Quick Trail Evaluation Form, these evaluations walk through the "whys" of evaluation in words in order to show the thought process. This valuable exercise not only evaluates the trails but also their underlying design. I've also briefly predicted the future of each trail.

With practice, the Foundation Level enables you to read trails like a book—clearly, concisely, consistently, and quickly. If you want to improve your trail design skills, I *strongly* encourage you to evaluate as many trails as you can—in writing if you're really serious—to practice working with the core concepts and learn about trail design in the process.

TRAIL EVALUATION
Southeast Utah desert

This several-mile-long loop trail is open to hiking only and receives relatively low use. It explores desert erosion features such as spires, small canyons, water- and wind-sculpted rock and mudstones.

Human perception
Natural shapes and edges abound in both site and trail, and the trail itself follows edges of rock outcrops, riding above and below cliff edges. In the left photo, there are gateways between two boulders just right of center, between two trees in lower right, and on a natural causeway behind the two trees at lower right. Sightlines expose constantly changing views.

Human feelings
Feeling of safety is enhanced by a berm on the outside edge of the tread (upper right photo). This is mostly because the soil doesn't hold outslope, but partly intentional to act as a curb at the tread edge. Efficiency is good—there's no obviously better route and no temptation to shortcut or bypass anything. It has lots of playfulness with changing grades, directions, tread surface (soil to rock), tight and varying relation to its site, frequently changing sightlines, and more. It continually pulls you along. Harmony is excellent since there isn't much we would want to change.

Physical forces and tread texture
Tread is primarily sandy silt that compacts to a reasonably firm and stable surface, yet displacement is relatively high for a hiking-only trail. It's difficult to maintain outslope since the fluffy soil sinks a lot with compaction, people avoid the outside edge, displacement forms a berm on the outside edge, and the trail is not frequently maintained. Tread soil erodes easily.

Tread watersheds
Although rain is rare, it can be fast and heavy (the entire canyon landscape clearly resulted from erosion). Tread watershed size can be large since slopes continue far above the photo area. Watershed slopes are steep. Runoff potential is extreme on steep, rocky, bare slopes with high runoff speed. With no tree canopy, splash erosion is high. Rain, snow, and their runoff are the sole water sources.

The trail accommodates these extreme conditions by carefully limiting tread lengths and tread width. Dip sustainability is good—note the gentle roller-coaster grade of most of the trail as it dips through natural drainages to let water cross quickly. In the usual sandy silt tread, grades are kept shallow. Steeper grades in the middle of the "S" curve are on solid rock with virtually no erosion risk.

Future prediction
In this context, the photographed section will serve for many years with little to no maintenance.

TRAIL EVALUATION
Alaska, near Anchorage

In a coniferous forest with rainforest-like climate, this site is cool and wet in the summer and frozen through a long winter. The trail is open to hiking, mountain biking, and horse use and has low to moderate use. This trail segment was constructed only a few years before the photo.

Human perception
Trees, plus the tread bench itself cut into the steep slope, anchor the trail. This segment is relatively straight but its edges have softened through weathering, giving them a more natural shape. Because the slope is so steep, the trail itself is a strong edge.

Human feelings
Our perception of safety is increased by the wide tread and by trees below the trail that make the slope feel less steep. A few roots and rocks are starting to emerge after compaction and displacement lowered the tread, but these aren't enough to make the tread feel inefficient. Rough (naturally shaped) edges and quick-growing moss help improve both playfulness and harmony.

Physical forces and tread texture
Tread is primarily gravelly loam high in clay and silt in varying proportions. As the protruding roots and rocks show, the tread has compacted several inches since construction. The tread has significant outslope although some rutting and tread erosion have occurred. Currently, the tread has good resistance to displacement and erosion, even on its steeper portions.

Tread watersheds
Tread watersheds are high but not long. Although it's not readily apparent in the photo, there's a dip on this tread segment just before it curves out of sight at the upper end. Watershed slope is steep. Runoff potential is moderate since forest litter is not that deep and the abundant moisture runs off of clay soil. Tree canopy protects some of the tread from splash erosion. Water comes in the form of abundant rain and runoff. No visible seeps or other water sources are in this particular tread watershed.

Currently, most tread drainage is through outslope. Since the tread has relatively low use, experiences relatively low displacement, and has some gravel content to help improve traction and structure in muddy conditions, outslope has a good chance of continuing to work here. With steep tread grades, though, displacement and erosion forces are both relatively high, potentially causing outslope to fail. But even if outslope does fail, enough tread watersheds were built into the alignment that the trail could drain exclusively through its dips without experiencing too much erosion.

Capsule evaluation
In terms of both structure and trail experience, it's working and I expect it to continue to function well. I wouldn't change anything.

TRAIL EVALUATION
Central Colorado Rockies

Open for all non-motorized uses, this is part of a destination trail in a subalpine forest.

Human perception
The trail is anchored by the trees but has no special relationships to them. In other words, it just passes by the trees. Natural shapes of rocks in the trail are the most apparent, but this is an unnatural concentration and not pleasant.

Human feelings
Since the tread is an obstacle course, safety and efficiency are poor. Note the new tread forming on the left as visitors seek a more efficient path. There's nothing playful or harmonious in the trail itself.

Physical forces and tread texture
The top soil layer is about 8 inches of forest litter. At this high, cool elevation, forest litter decomposes slowly, creating a deep, spongy top layer that readily compacts, displaces, and erodes. The initial tread was compacted and displaced into a rut by use. Erosion did the rest. The tread has now eroded down to a sandy subsoil that is also erosive.

Tread watersheds
Tread watershed size is much too large because crests and dips are many hundreds of feet apart. Although watershed slope and tread grade are not that steep, runoff potential is moderate, splash erosion is slightly reduced by tree canopy, trail use is low, and the tread was originally narrow, the large tread watershed delivers too much water to erosive tread. Occasional heavy rains, and frequent moderate and light rains, increased erosion. Contributions of water sources other than rain and surface runoff are unknown, but it's likely that the deep tread tapped and drained a seasonally high water table.

Dip sustainability is very poor. Once the tread compacted and displaced, it undercut any dips it may have had, merging any smaller tread watersheds it may have had into one excessively large tread watershed. The deep rut is difficult to drain on these shallow slopes. Note the desperate waterbar in the center foreground with stones that look like a dam in the trail and a drainage channel to the right. This is too little, too late. It now takes major effort and long, inadequately sloped drainage channels to try to drain this deep tread.

Future prediction
Even if new waterbars were added, the new tread forming on the left will become deeper with compaction and displacement. Yet it probably won't erode much since it would drain into the lower existing tread that will function like a ditch.

Beyond adding waterbars, the trail could be rerouted and the old tread closed and restored, or firm fill could be imported to raise the tread above the existing ground level. In the latter, the raised tread would need periodic dips to form smaller tread watersheds and cross drains.

TRAIL EVALUATION

Eastern slope of the Sierra Nevada, California

In federally designated wilderness, this heavily used trail is open only to hikers and equestrians. The trail has heavy horse use, including pack trains such as shown here. This evaluation is for the trail in this type of environment rather than just the photographed segments.

Human perception
The landscape is the star here. Although these photos don't show it, the nearby mountains and lakes—the crest of the High Sierra—steal the show. Natural shapes, anchors, and gateways abound, and the rugged site forces the trail to conform to it.

Human feelings
The feeling of safety is about average for mountain trails. Efficiency varies, mostly because the tread is sometimes so loose that walking can be difficult, yet the trail doesn't encourage shortcutting. Playfulness is everywhere because the trail has to conform to the site, and harmony ensues where the trail is most tightly woven into (anchored by) the site.

Physical forces and tread texture
All tread material is decomposed granite, a loose mix of silt, sand, and gravelly particles (mostly quartz) that don't bind with anything. Here, particles have rounded edges and almost no binding (no clay), reducing mechanical and electrical stabilization and making treads loose and unstable. Outslope is absolutely unfeasible with such easily displaced tread. On the plus side, the tread is so porous that only major rain and runoff can put enough water on the tread to erode it. Compaction hardly sinks the gravelly tread.

Tread watersheds
The trail keeps its tread watersheds small through careful alignment and extensive construction of waterbars, drainage dips, and other drainage techniques. Watershed slopes are generally steep, but runoff potential is proportional to the amount of solid rock. Runoff is very high in the left photo. Splash erosion potential is high, but the gravelly tread helps resist it. Occasional heavy rains do provide enough water to cause erosion. Tread width is about as narrow as it can be considering the usage.

Much of the problem, though, is easily displaced tread with high displacement usage (horses). Grades are kept low and tread lengths are kept short to limit erosion. To help reduce tread grades, risers were added where other treads wouldn't need them (right photo). With low grades and frequent dips formed through both construction and alignment, dip sustainability is good overall.

Future prediction
Material is constantly kicked out of the tread through displacement, but good dip sustainability and frequent maintenance should enable this rewarding trail to successfully work within difficult constraints.

TRAIL EVALUATION
ATV trail, central Minnesota

This dedicated ATV trail is only about a year old. It was planned but the tread was "ridden in" rather than constructed, so no excavation was performed.

Human perception
The trail casually winds through a pleasant, brushy, previously logged hardwood forest of widely varying slopes of glacial origin. Although there are no strong anchors or edges, individual trees anchor the trail and form occasional gateways in the relatively undifferentiated forest. The trail, however, would be more anchored in the site if it traversed instead of avoided the steeper slopes.

Human feelings
Safety is okay. Efficiency is okay but decreases when the trail is wet or muddy. Playfulness comes from its natural shape as the trail wraps around trees and varying slopes. Harmony comes from its natural shape and narrow tread but diminishes where there's excess tread displacement (right photo).

Physical forces and tread texture
Tread is loamy with highly varying amounts of sand and silt from glacial deposits, with a thin layer of organic soil on top. Because ATVs exert high displacement forces but low compaction, the tread tends to be kicked out by the tires. Both photos show compaction and displacement in the tread, especially under the tires. This is expected. In the right photo, muddy ruts were filled with large washed gravel, but because it had no fine particles, it was promptly displaced out of the tread and to the outside of the curves. Steeper trail grades erode readily, especially in the more sandy soils.

Tread watersheds
With an irregular glacial topography of varying slopes, low areas, and nearly level areas, tread watershed size and slope varies widely. Runoff potential is moderate since the organic layer is not deep and the soil seems poorly or only moderately drained. Tree canopy reduces splash erosion. Intense or long-duration rains are common. Rainwater collects in the low areas, and there may be a seasonally high water table.

Tread width is as narrow as possible. Tread grades and tread lengths vary, with fewer problems where the tread grades are about 4% to 8% and relatively short in length (left photo). Steeper grades readily erode, grades less than 4% don't drain quickly enough, and longer tread lengths accumulate too much water in the tread ruts.

The main problem is dip sustainability, specifically lack of quick drainage to the side and poor tread resistance to ATV displacement. Tread on pronounced slopes and with pronounced but not excessive grades performs well (left photo). Tread in more level areas or that follows fall lines accumulates water in the sunken tread, creating mud, accelerating displacement, and/or requiring tread hardening (right photo).

Future prediction
The trail should have traversed steeper slopes and avoided level areas wherever possible. This would maximize dip sustainability, helping solve one of the largest problems. Reroutes would help—many problems were on near-level ground adjacent to a slope that would have been a better location.

TRAIL EVALUATION
Wet meadow in northern Cascades, Washington

Note: This method of accommodating water is only appropriate in particular contexts. Book 2 discusses many methods.

This wet meadow has high annual precipitation and a high water table. The trail, which receives low to moderate use, is open to hikers and equestrians.

Human perception
Natural shapes in the trail and ditches echo abundant natural shapes in the site. Trees and shrubs of various sizes, plus grasses between the tread and ditch in places, help anchor the trail. Ditches, especially where water is visible, create new edges.

Human feelings
Tread is sufficiently wide and stable to feel safe and be efficient. Despite the obvious manipulation, playfulness and harmony come from using natural shapes (avoiding all straight lines); working around site anchors; hiding manufactured materials; and relaxing the geometry, widths, and spacing of tread and ditches.

Physical forces and tread texture
Gravelly loam tread in this volcanic region can be stable when wet—but not when saturated. Lowering the water table, draining tread surface water, and compacting the tread surface through use sufficiently stabilizes the tread. Gravel content greatly helps strengthen wet tread and reduces displacement, erosion, and rutting from compaction. The tread itself, however, is not that hard and erodes if subject to fast-moving water.

Tread watersheds
Tread watersheds are kept small. Relatively deep, wide ditches with short runs frequently drain away from the trail (to the left) so that even heavy rains can be carried without overtopping ditches and culverts. Stones at the bottom of the photo are the tailwall of an angled culvert that follows the fall line to drain quickly, minimizing clogging.

Watershed slope is gentle but has high runoff potential due to abundant rainfall, high water table, and gravel content. Tread width is about as narrow as it can be for the type and amount of use, and extra width is needed next to open ditches.

Water sources are direct rainfall, surface runoff, and subsurface seeping that continues through much of the year. Since this slope is not technically considered a wetland, ditches can be used to lower the local water table, slightly modifying site hydrology but with minimal site impact. Because the tread is sufficiently stable when well drained, keeping the top 8 inches of tread drained is sufficient for sustainable trail use.

With this tread design, dip sustainability concerns the ability of ditches and culverts to function like dips, and here they indeed have all six characteristics of sustainable dips. The tread surface, however, has no dips, so technically it's a long tread watershed consisting mostly of itself. It uses culverts instead of open dips and relies on outslope, inslope, and crowning to pitch tread surface water toward the ditches. Hence maintaining tread shape is **critical** for draining the tread surface. Low to moderate tread grades help reduce excess tread shape change from displacement and erosion.

Future prediction
The trail can remain sustainable as long as tread surface shape (outslope, inslope, crowning) is maintained and ditches and culverts are kept clear.

TRAIL EVALUATION

Historic 1855 pack trail, northern California

This is on the South Kelsey Trail, Smith River National Recreation Area, Six Rivers National Forest, CA. From the USFS trail brochure:

> From 1855-1880, the historic Kelsey Trail was the major transportation link between Crescent City and mines in the Klamath River region and Yreka. The people of Yreka raised money to build the trail from Yreka across the Marble Mountains to the Klamath and the people of Crescent City paid for their section from Crescent City to the Klamath… Mule trains carried supplies inland and, presumably, gold and other minerals back to Crescent City [on the coast] and eventually San Francisco [by ship].

The singletrack trail is open to hikers and horses.

Human perception
The trail is well anchored by being cut into very steep, forested hillsides above a whitewater river, making the trail a strong edge in itself. Stone retaining walls (foreground) support many parts of the trail, enhancing the shape, drama, and feeling of the edge and creating multiple gateways. The river is visible at times through gateway-like openings.

Human feelings
For safety, a nearly continuous curb-like berm on the outside edge helps keep feet and hooves on the tread, and thick brush and forest below the tread makes the dropoff feel less steep than it is. Efficiency is excellent with no tread obstacles, no risers or steps, no difficult-to-negotiate drainage devices, and very little erosion. Playfulness comes from the pioneer gold rush spirit in which the trail was built—narrow width, rough retaining walls, form following function, built by people who knew how soil and rock behave. Harmony comes from good trail condition and 150 years of moss and lush growth that covers all construction scars.

Physical forces and tread texture
Tread is clay (rock-hard when dry, gumbo when wet) surfaced with limestone chips (gravel) that trail use has compacted into the top layer of clay. Twenty-five years of mule trains laden with gold and supplies firmly compacted the surface until no displacement and little erosion occurs today.

Tread watersheds
The height of the slopes above the trail make large tread watersheds, and watershed slopes are steep to extreme. (Whoever laid out the route must have had a hell of a tough time.) A moderately thick forest canopy and a relatively thin layer of forest litter over steep clay and rocky slopes creates moderate to high runoff potential. Many, but not all, parts of the tread are protected from splash erosion. Water sources include coastal fog, slow but light rains, and occasional heavy rains.

As the photo shows, the hardened, compacted tread has no outslope. The rut from mule trains is still clearly visible, but the clay-gravel tread is now so hard that current, light trail use has little effect.

Tread grades were kept as low as practical for the mule trains. Tread lengths are reasonable even though the trail continually climbs and drops hundreds of feet while traversing the canyon wall. The tread drains entirely through its dips and through a few outflows cut through the outside berm. Since most dips are formed by alignment, drain quickly to the side, and have stable tread, dips still function after 150 years—many of those years with no maintenance and even some forest fires.

Future prediction
There's no reason why this trail can't last for many future centuries.

TRAIL EVALUATION
Mountain bike trail, southeastern Pennsylvania

This unofficial, unmapped trail recently formed entirely through informal mountain bike use in an open forest. Because the trail has no destination and exists purely for the fun of being on it, it has a freedom that destination trails don't. Like most visitor-formed trails, it has its problems—but it also has many successful implementations of the core concepts.

Human perception
Natural shapes become curvilinear to fit the turning ability of bikes, yet still retain an unpredictable, opportunistic quirkiness. Like many trails, it wraps itself around anchors in some places (left photo) while not paying too much attention to them in others (right photo). Some of its most interesting parts are where it plays with anchors and edges (both photos).

Human feelings
Those who formed this trail were seeking more risk than nearby trails offer, so it intentionally has tight clearance, obstacles to negotiate, and occasional fast cruiser sections (right). Efficiency is good since people went where they wanted in the first place. Playfulness is the point, and harmony comes from having relatively little impact (at least for now).

Physical forces and tread texture
Tread is clay loam with moderate resistance to displacement and erosion, and moderate tendency to form a rut through compaction and displacement. This type of soil tends to be hard and firm when dry but plastic when wet and muddy if saturated. Continued use will deepen the tread. Note the superelevated curve forming in the right photo.

Tread watersheds
In some ways, the site supports the trail. Bicyclists' preference for frequently going up and down creates crests and dips that aid drainage (right photo). Tread watershed slopes are often not that steep or long although runoff potential is high and heavy rains occur often. Tree canopy reduces splash erosion. Tread width is narrow, reducing the surface area for erosion. There appear to be no water sources other than precipitation and a few defined drainages. Trail use is light at this point and limited to bikes. As noted above, tread texture tends to be firm when dry.

Tread grade, tread length, and dip sustainability, however, were not considered. In time, that will be the downfall of many sections as poor drainage, erosion, and puddles become more pronounced than they are on this very young trail.

Future prediction
Many parts of the trail will develop drainage problems as the rut deepens, outside curves superelevate, and more storms occur. It would have been possible, though, to design a trail that was almost as much fun yet much more sustainable.

TRAIL EVALUATION
Crushed stone trail tread, Colorado

Of the trails evaluated in this chapter, this is the only one that I designed. On the National Center for Atmospheric Research site in Boulder, it's both an interpretive loop trail and a popular trailhead to the mountains in the backgrounds of both photos. Trail use is heavy with scores of hikers per day and many hundreds on weekends. The crushed stone tread is barrier-free for wheelchair users.

Human perception
Perched on a narrow mesa top with meadows and pines, one leg of the loop follows the edge of the mesa and features several overlooks (left photo). Trees, rocks, rock formations, and the mesa edge thoroughly anchor this winding leg. The other leg (right photo) is straighter and more direct for the majority of visitors who are just passing through to the mountains beyond, yet it, too, incorporates natural shapes that anchor it within the meadows.

Human feelings
Overlooks built on retaining walls have native stone curbs (the top of the wall), but no other safety features were added for wheelchair users. Efficiency is pretty good although corner cutting has occurred at some sharp corners. Playfulness comes from varying tread width, wrapping around or coming close to trees and rocks (incorporated anchors), and abundant natural shapes in the trail itself. Harmony comes from playfulness, going where people want to go (overlooks and along the edge), and frequent anchoring. To match native onsite stone, the orange-pink crushed stone in the tread was made from the same type of stone that surrounds the trail and is in the rock formations in the background.

Physical forces and tread texture
Previous native soil treads had widened up to 14' in width by visitors attempting to avoid mud, especially during winter and spring. The crushed stone tread doesn't sink with compaction and suffers relatively little displacement. However, 15 years of heavy and ever-increasing use, occasional torrential rains, and mountain winds exceeding 100 mph have caused displacement and erosion. The original crown of the trail has been lost in many places and the tread surface has lost much of its dust and fine particles.

Tread watersheds
The peninsula-like mesa top climbs at 4% to 11% and travel patterns dictate that the trail climb with it, often directly along or very close to the fall line. Because of heavy rains, high runoff potential, high splash erosion, and large tread watersheds in places, the tread was crowned 2-4". For added drainage, 52 drainage dips were built into the tread with closer spacing in the steeper sections.

Now, 15 years later, many of the dips have filled in from displacement and lack of quick drainage to the side on near-fall line alignments. Much of the crown has been lost to displacement, erosion, and poor maintenance (lack of maintenance, no dip cleanout, and skimming the crown to fill a washout caused by loss of dips). Nonetheless, the trail still functions well after 15 years of year-round use.

Designer's comments
Knowing what I know now, I would have done a few things a bit differently, but it's a largely successful combination of hardening the tread and shaping a naturalistic trail experience.

● **CHAPTER 9**

Trails by Design

Using the Foundation Level

Even without knowing the details of trail shaping and management at the Middle and Upper Levels, the Foundation Level is a powerful tool. It enables you to understand, recognize, and explain the basic human and physical relationships that underlie each and every natural surface trail. And because techniques that are fully supported by the Foundation Level function with minimal ongoing effort, all sustainable natural surface trail design, construction, maintenance, and management begins here.

Here's how to use the Foundation Level to best advantage:

Put a new foundation under what you already know
The Foundation Level underlies traditional trail design and construction. Make room for new concepts and allow the system to unfurl as you put a new foundation under what you already know.

Use the Foundation Level on all natural surface trails
Each of the eleven concepts applies to all natural surface trails regardless of trail type, use, or location. *There are no exceptions.* You must consider the entire Foundation Level for any trail or trail segment.

Let the Foundation Level generate trails*
By thoroughly describing the context, the Foundation Level shows us where to look, enables us to clearly see what's actually happening where the trail touches the site, helps us know *why* it's happening, and leads us to the implications. In the process of examining your context in terms of all eleven concepts, you'll be guided toward workable situations—and away from unworkable ones—as it literally guides your thoughts and generates sustainable trails in flexible, logical, and creative ways. This is the major advantage of a system of thought rather than a system of rules. If you're like most people, you'll find yourself working with trails in refreshing new ways.

Use its language, pictures, and names with yourself and others
The Foundation Level intentionally uses words and pictures as powerful design tools. They guide our thinking and invoke whole fields of meaning for ourselves and others quickly, fluidly, and easily—simply by thinking, saying the word, or seeing the picture.

Names—even more powerful than words—act like pictures to invoke even larger fields of words, pictures, feelings, and nuances as well as the relations between them. Names let us express complex concepts faster and easier than words, and in ways not even remotely possible with rules. *That's why it's essential that basic forces and*

* Note that trail shaping and management at the Middle and Upper Levels often affect Foundation Level decisions (see pages 77-78). Nonetheless, even by itself, the Foundation Level generates trails that are sustainable within its own context.

relationships be named. And it's just as essential to use those names as "building blocks" to form and communicate even more complex ideas until we can clearly understand the entire context of any trail situation.*

Consider the human as well as the physical aspects of trails and sites

In traditional trail design, construction, and maintenance, there's strong emphasis on physical aspects yet relatively little consideration of human aspects. That can result in relatively unsatisfying recreational trails, or, at worst, trails that fail because their human aspects were not appropriately considered. The Foundation Level, however, explicitly relates human and physical aspects in many ways. Six of its eleven concepts are primarily human aspects. Consciously use those human aspects to help shape trails that are both sustainable *and* enjoyable.

Expect tread to change shape over time and always predict the future

As you've seen, the concept and mechanics of how trail tread changes shape over time are built into the Foundation Level. Natural surface trail treads are dynamic systems that *will* predictably change over time. Understanding the process of change is essential for making predictions on which to base sustainable design. If you're not expecting tread shape change or you're not using the Foundation Level to predict the future of trail treads, you're ignoring one of its most critical and unique aspects.

Consider different trail types as instances of the same basic forces and relationships

Many people think exclusively about, or in terms of, particular trail types for particular modalities, such as horse trails or ATV trails. Or they only think about trails in limited locations or ecosystems. The Foundation Level creates a far more rewarding possibility: considering all different trail types and locations as simply different degrees of the same forces and different instances of the same types of relationships. For example, a rocky horse trail in Kentucky, a hiking trail in the wet Cascades of Washington; a crushed stone wheelchair-accessible trail in Michigan; an ATV trail in a broad Alaskan valley; a mountain bike trail in the California foothills; a motorcycle trail in Missouri; and a hike/bike/horse trail in Virginia are all are just different degrees of the same forces and different instances of the same types of relationships.

---

* I know it can be difficult for men and women with a "manly" attitude about trails to professionally think about—let alone talk or *(gasp!)* write about—playfulness, harmony, or feelings. Yet the human aspects are every bit as important as the physical ones, recreational trails are about feelings, and resisting the use of human terms and feelings is largely how we avoided seeing this simple system long ago. So, for you engineers, bureaucrats, rangers, and tough trailbuilders—**soften up!** Try it, talk your colleagues into trying it, too, and see how it enriches your professional and personal trail life.

The Foundation Level in a Nutshell

For any trail, the eleven core concepts show us the essential forces and relationships, how they act, and how they interact. Natural shapes give us the shape and character of naturalistic trails. Anchors, edges, and gateways help create distinct trail experiences while tightly relating the trail to the site.

Safety considers our perception of safety, how the trail creates it, and how that affects us. Efficiency guides us to shape trails that are easier to use than bypass. Playfulness reminds us of the joys of adaptations to the exact context, of quirks and relaxed geometry, and of trail aspects that aren't completely utilitarian. Harmony—having everything working together, and feeling whole as a result—is our overall goal.

Compaction, displacement, and erosion explain and predict the inevitable human and natural forces acting on all natural surface trail treads, as well as their interactions and consequences.

Tread texture provides easy-to-use tools to explain and predict tread behavior and characteristics based on the characteristics of soils and tread materials.

And tread watersheds unite the physical forces, tread texture, water sources, and water movement onto and off of the tread, enabling us to see and predict what's likely to be sustainable against erosion—as well as what won't be as sustainable.

Note that we're not trying to make all trails "the same." The whole idea of the Foundation Level is to treat different trails differently to suit their unique individual contexts. It's easier to do that, though, when we clearly view each trail as a specific instance based on Foundation Level forces and relationships.

From this higher point of view, it's easier to consider and appropriately optimize each trail type for its use and location since many aspects that apply to one trail type also apply to another. It's also easier to appreciate each trail for what it is and to realize its potential. Not only does this make you a better trail designer, but it also tends to increase your understanding (and tolerance, if you have a bias) of different trail types and different modalities.

Appropriately incorporate new modalities, materials, and techniques
New modalities, new tread surfaces, new materials, and new trail shaping techniques are merely new instances supported by the same familiar forces and relationships. You can learn how they fit into the Foundation Level. And by thoroughly studying new specifics and ideas and their implications, you can usually determine what will or won't work before having to try them in the field.

Continually work to engender stewardship for natural resources
Engendering stewardship is an extension of sustainability. If people feel a sense of stewardship, they'll help sustain the context. Hence part of every decision should be how the results can increase appreciation and respect for the trail, site, and natural resources, and thereby help engender stewardship for same.

Appreciation, respect, and stewardship are generated, or earned, by focusing on and working with nature—by tightly weaving trails into sites using natural shapes and anchors; by arranging trails to generate "positive" human feelings; by accommodating compaction, displacement, and erosion instead of fighting them; and by using tread watersheds to naturalistically limit water-related problems. All of these help form harmonious, rewarding, and sustainable trail treads that celebrate and incorporate nature rather than ignore or overengineer it.

By design, engendering stewardship is intrinsic in the system. Yet it's up to you and the trail to fully manifest it in the world.

The Next Two Levels
While the Foundation Level tells you a great deal about how natural surface trail treads work and what a sustainable tread needs, it doesn't tell you how to fully design a trail. That's a larger process requiring the Middle and Upper Levels.

Middle Level: Trail Shaping Techniques
Book 2 in the Trails by Design series, *Shaping Natural Surface Trails by Design: Key Patterns for Forming Sustainable, Enjoyable Trails*, explores key aspects of the Middle Level. Specific trail design, construction, and maintenance techniques—more accurately called

It should now be obvious why the Foundation Level is depicted as the tip of the cone. Its eleven core concepts literally support all the myriad trail shaping (design, construction, maintenance) techniques above it, as well as trail purpose and management.

shaping techniques—can be concisely represented as "patterns" based on the Foundation Level. Patterns are a powerful way to concisely specify complex shaping information. Most trail shaping techniques, including known and trusted techniques, can be improved and easily learned as instances of a handful of key patterns.

Book 2 shows how to optimize trail shaping for maximum sustainability and visitor enjoyment for any trail use. It emphasizes ways to minimize effort in both initial trail shaping and ongoing maintenance (reshaping). Where appropriate in your context, they can result in highly naturalistic and enjoyable trails that are more sustainable and less expensive than traditionally constructed trails. Book 2 also discusses low-tech, low-cost ways to accommodate high-displacement modalities as well as low- and high-tech tread hardening methods.

Upper Level: Trail Purpose and Management

Book 3 in the series, *Managing Trails by Design: Integrating Stewardship, Sustainability and the Trail Experience,* discusses the Upper Level of Trail Purpose and Management. It includes recreational motivations of visitors, human psychology, reasons to have (or not have) a trail, ecological considerations and impacts, overall trail planning and siting, types and amount of trail use, visitor conflicts, management techniques, maximizing and managing the visitor's trail experience, and much more. These are all discussed in terms of the Foundation Level and the techniques of the Middle Level, especially those introduced in Book 2.

For more information on these books, trainings and workshops by Natureshape, or to order books and publications, please visit **natureshape.com** or see the last page of this book.

About the Author

Troy Scott Parker is the president of Natureshape LLC. He has designed and built trails for the National Park Service, USDA Forest Service, The Nature Conservancy, and others. He authored the popular *Trails Design and Management Handbook* for the Open Space & Trails Department of Pitkin County, Colorado, an internationally used design guide for multiple use concrete/asphalt trails, crushed stone trails, boardwalks, and other trail features. For the Minnesota Department of Natural Resources, he authored a major portion of *Site-Level Design and Development Guidelines for Recreational Trails,* a comprehensive planning, design, construction, and maintenance guide for all trail types and uses (unpublished as of this writing). He is a past president of the Professional Trailbuilders Association and a popular presenter at trail conferences.

Involved in trail research and education since 1985, he has a growing collection of more than 13,000 photographs of all aspects of natural surface and paved trails, trail bridges, boardwalks, and other trail-related features in over 100 categories. All photos in this book are from that collection.

From other design-oriented careers, he is also an experienced book designer, graphic artist, and software engineer.

Natureshape LLC provides a wide range of trail-related
training, publishing, consulting, trail design, and information resources

Training

- Workshops (indoors, outdoors, or both) on trail shaping—design, construction, maintenance
- Custom field trainings at your location
- Conference presentations

Publishing

- Books and publications on trail design, construction, maintenance, and/or management for any trail type
- Design, layout, graphic design, photo scanning and prepress, and typesetting for books and publications; and website design
- Potential authors of trail-related "how-to" books are encouraged to contact us

Consulting

- Consulting to improve existing or proposed trails, site integration, trail features, site restoration, and more
- Custom publications on trail design, construction, maintenance, and management

Trail design

- Sustainable, naturalistic trail design for all types of trails and uses
- Emphasis on using appropriate design to provide enjoyable trail experiences, maximize visitor appreciation and respect for the site, and minimize construction and maintenance costs

Information resources for sustainable, enjoyable trails

- We sincerely want you to have the best information on the art and science of sustainable, enjoyable trails—*even if we didn't publish it.* To help you find the published trail information you need from any source, **natureshape.com** offers in-depth reviews and comparisons of current, high-quality, trail-related books and publications authored by federal, state, and local governments, non-profits, and for-profits.

For more information, please visit natureshape.com or contact
Natureshape LLC

| 8285 Kincross Drive | natureshape.com | Ph (303) 530-1785 |
| Boulder, CO 80301 | info@natureshape.com | Fax (303) 530-4757 |

Order Form

Natureshape LLC
8285 Kincross Drive • Boulder, CO 80301-4228 • USA
Ph (303) 530-1785 • Fax (303) 530-4757 • orders@natureshape.com
natureshape.com

*For potential discounts, fastest service, and all international orders, please order online at **natureshape.com***

Name

Agency or organization

Address

City State ZIP

Phone (just in case)

ITEM	QTY.	ITEM PRICE	TOTAL
Natural Surface Trails by Design: Physical and Human Design Essentials of Sustainable, Enjoyable Trails		$36.00	
Shaping Natural Surface Trails by Design: Key Patterns for Forming Sustainable, Enjoyable Trails		contact us	
Managing Trails by Design: Integrating Stewardship, Sustainability and the Trail Experience		contact us	
		SUBTOTAL	
		SHIPPING & HANDLING CHARGE (see below)	
		ADD 4.25% SALES TAX TO SUBTOTAL **PLUS** SHIPPING if shipped to a Colorado address	
Thank you for your order!		**TOTAL**	

METHOD OF PAYMENT

Enclose your check or credit card information as indicated below.

❏ Check enclosed   ❏ VISA   ❏ MasterCard   ❏ Discover

Card account number:

Signature of Authorized Buyer

Expiration date on card: ☐☐ / ☐☐
Month   Year

SHIPPING OPTIONS AND CHARGES

Choose Standard (US Mail or UPS) shipping or provide your FedEx account number and preferred delivery time ($3.00 handling fee)

❏ **Standard shipping** (US Mail or UPS Ground)
Add $5.00 for the first book and $0.50 for each additional book

❏ **FedEx shipping billed directly to your FedEx account**
NOTE: Enter $3.00 for handling on form above
Your FedEx acct #: _____
❏ Priority Overnight   ❏ Standard Overnight
❏ Second Business Day   ❏ Express Saver (3rd business day)